Haline Rocha
Colombo Tassinari

Geological Study on Unconventional Natural Gas in Brazil

Haline Rocha
Colombo Tassinari

Geological Study on Unconventional Natural Gas in Brazil

Evaluation of the Potential for CNG Generation in the Paraná Basin and the Feasibility of End Use in Thermoelectric Generation

ScienciaScripts

Cover image: www.ingimage.com

This book is a translation from the original published under ISBN 978-3-330-76153-7.

Publisher:
Sciencia Scripts
is a trademark of
Dodo Books Indian Ocean Ltd. and OmniScriptum S.R.L publishing group

120 High Road, East Finchley, London, N2 9ED, United Kingdom
Str. Armeneasca 28/1, office 1, Chisinau MD-2012, Republic of Moldova, Europe
Managing Directors: Ieva Konstantinova, Victoria Ursu
info@omniscriptum.com

Printed at: see last page
ISBN: 978-620-8-37178-4

ACKNOWLEDGEMENTS

My sincere thanks go to everyone who contributed directly and indirectly to the development and completion of this work.

To the University of São Paulo, Institute of Energy and Environment and Postgraduate Programme in Energy, for their institutional support.

To the Coordination for the Improvement of Higher Education Personnel (Capes), for the financial support.

To my teachers and friends Edmilson Moutinho dos Santos, Hirdan Katarina de Medeiros and Estanislau Luczynski for their support and friendship.

To my dear supervisor Colombo Celso Gaeta Tassinari for his knowledge and dedication.

To my family, for everything.

SUMMARY

This master's thesis consists of a geological study of the potential for exploration and production of unconventional natural gas in the Paraná Basin, in the south and southeast of Brazil and northern Uruguay. To this end, the Irati and Ponta Grossa geological formations, located in the southern portion of the basin and characterised by their hydrocarbon generating potential, were analysed. The study also includes an assessment of the feasibility of supplying the Uruguaiana thermoelectric plant with this resource, taking into account the associated geological, socioeconomic and environmental aspects. The project aims to encourage the study of unconventional energy resources, with emphasis on shale gas, the main source of unconventional natural gas explored in the world, as well as highlighting the benefits of the growing implementation of natural gas in the Brazilian energy matrix, both in strategic and economic terms. The project proposes the development of a descriptive-comparative model based on the geological aspects of the main shale gas producing fields in the United States, the world's leading producer, through which the potential for the occurrence of this resource in the black shales corresponding to the Irati and Ponta Grossa Formations is assessed, according to geochemical, petrophysical and geological parameters present in proven producing formations in the USA: Marcellus, Barnett and Eagle Ford. The purpose of the descriptive model is to assess the potential for exploration, production and distribution of unconventional natural gas in the south and southeast regions of Brazil, which represent the largest consumer market and demand for energy in the country. Based on the comparative analysis carried out, the Brazilian formations are considered to have moderate to high potential for shale gas generation. However, the feasibility of supplying the Uruguaiana thermoelectric plant depends, in addition to the local geology, on political and regulatory factors that are still not well established in the country to encourage the production of unconventional energy resources.

Keywords: Unconventional Natural Gas, Shale Gas, Irati and Ponta Grossa Formations, Brazilian Energy Matrix.

SUMMARY

CHAPTER 1

INTRODUCTION

The constant search for energy resources has resulted in several advances in the unconventional hydrocarbons sector. Research carried out by the U.S. Geological Survey (USGS) found that natural gas could not only be exploited in reservoirs with a conventional geological configuration, such as anticlinal traps, but could also be found in rocks with low permeability: a property that allows the same rock package to play the role of generator, reservoir and sealant. The discovery revolutionised the US oil industry which, by developing suitable exploration techniques, made it economically viable to exploit this resource.

The main exploration techniques for unconventional systems were developed in the United States from studies carried out in the Bakken formation, in the Williston sedimentary basin in North Dakota, and are called hydraulic fracturing and horizontal drilling. The first term refers to the hydraulic fracturing technique, which has been used since the 1940s and basically consists of injecting a large volume of water under high pressure, together with sands and chemical compounds, into an oil-saturated formation with the aim of fracturing the rock and generating a passage route for the explored fluid. Another exploration technique, called horizontal drilling, emerged in the 1980s when technological advances allowed operators to drill to certain depths and then drill at an angle, or even completely horizontally. This technique allowed greater access to the reservoir, resulting in greater hydrocarbon recovery. Based on this knowledge, companies began to combine both drilling techniques, bringing innovation to the market and lucrative returns. However, this economic benefit can be associated with high environmental costs: since the technique of exploiting this resource includes the injection of millions of litres of water usually associated with chemical compounds that could potentially result in the contamination of aquifers, it is necessary to assess and prevent these possible environmental impacts (Yergin, 2011).

The term shale gas corresponds to deposits of natural gas that are trapped in pelitic formations rich in organic matter (black shales) due to the possible overpressure stage in which they are found. Shales are generally characterised by their low permeability, which makes it difficult to produce oil directly from this lithology; it was previously considered an uncommercial or uneconomic source of hydrocarbons. However, the reduced permeability of shales can act as a factor in retaining natural gas within this rock package, among the lamellae that accompany the foliation. Shales that contain significant amounts of natural gas have the following properties in common: composition enriched in organic matter (from 0.5 to 25 per cent) and generally present in the field of thermogenic gas (temperature range, between 120 and 180°C, low permeability: < 0.01 mD, and sufficiently high burial, converting oil into natural gas). The gas generated from this process can be contained in natural fractures in the rock package, inside pores, and/or adsorbed to the organic-sedimentary material. The gas retained in pores and fractures is recovered during the development and production of the oil system, while the adsorbed gas is released as the internal pressure of the reservoir decreases. Unconventional natural gas, which was previously not economically accessible, began to be significantly exploited with the development and application of the aforementioned exploration techniques, which allowed the direct production of oil from shales to become economically viable (Das Virgens, 2011).

The combination of hydraulic fracturing and horizontal drilling technologies innovated the US oil market and became known as the *Unconventional Gas Revolution* (Yergin, 2011). In the 2000s, shale gas production represented just 1% of total annual natural gas production in the US; by 2010, production had reached over 20% of the total, and according to the *Energy Information Administration* (EIA), production from this resource will reach 46% of total gas production by 2035. Currently, shale gas is the world's most exploited unconventional resource and accounts for more than 50% of US unconventional production (Figure 1). According to studies and estimates, natural gas production in the US in 2011 was calculated at 23.0 trillion cubic feet (Tcf[1]) and could reach

[1] 01 Tcf = 28.32 Bm3 or 01 cf = 0.0283 m3.

33.1 Tcf by 2040, which represents a 44% increase in production. Almost all of this increase in production is due to the exploitation of shale gas (Figure 17), which grew from 7.8 Tcf in 2011 to 16.7 Tcf in 2040 (EIA, 2012).

U.S. dry natural gas production trillion cubic feet

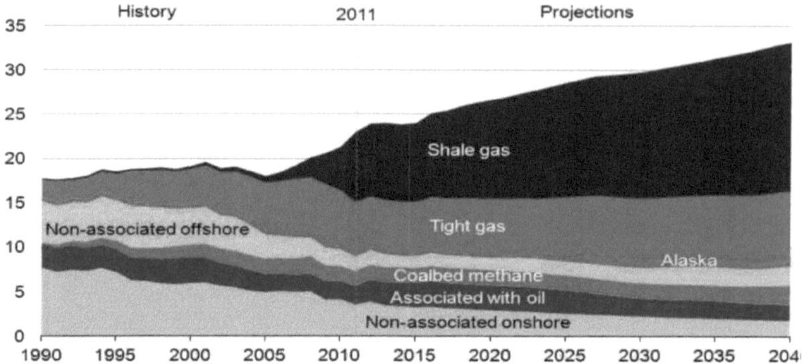

Figure 1 - Graphical representation of North American production of unconventional natural gas, indicating the types of reserves recovered and their respective production measured in trillions of cubic feet over time. Source: EIA (2012).

North American *shale gas* production began in the *Barnett* Formation in the state of Texas and intensified in the *Eagle Ford* Formation, also in Texas, and *the Marcellus* Formation, located in the Appalachian Mountains region in the eastern United States (EIA, 2012). Through surface and subsurface geophysical data, combined with geological studies, other reserves of this type of gas have been found throughout the country (Figure 2).

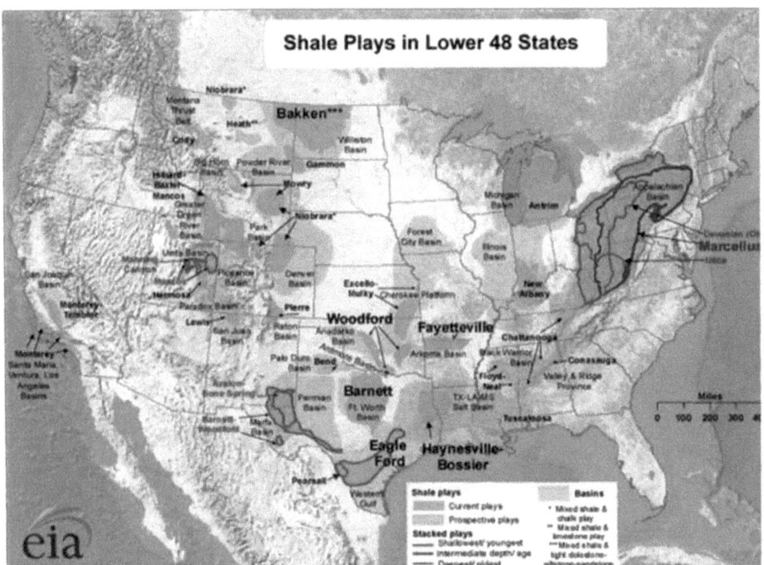

Figure 2 - Map of the United States showing the distribution of shale gas reserves. The sedimentary basins are identified as producing or with estimated potential. Source: EIA (2012).

Benefiting from this technological advance, natural gas has become competitively priced in the energy resources market, gaining advantages due to its lower pollutant content compared to its competitors: coal and oil, because it represents a lower environmental risk than other fossil fuels and nuclear energy, and also advantages over renewable energy sources such as wind power, due to its lower production costs.

According to the EIA (2012), shale gas deposits are widely distributed geographically (Figure 3), with the largest concentrations in China, with estimated values of 1,115 Tcf, followed by Argentina (802 Tcf), Algeria (707 Tcf), the United States (665 Tcf), and Canada (573 Tcf). Although several countries have geological potential for shale gas, until 2013, only the United States, Canada and China produced this resource on a commercial scale. Brazil is tenth on the list of countries with the greatest potential for this resource, estimated at 245 Tcf. However, only 14 national reserves have had their potential proven, a fact that reflects the insufficiency of national research in the unconventional hydrocarbons sector.

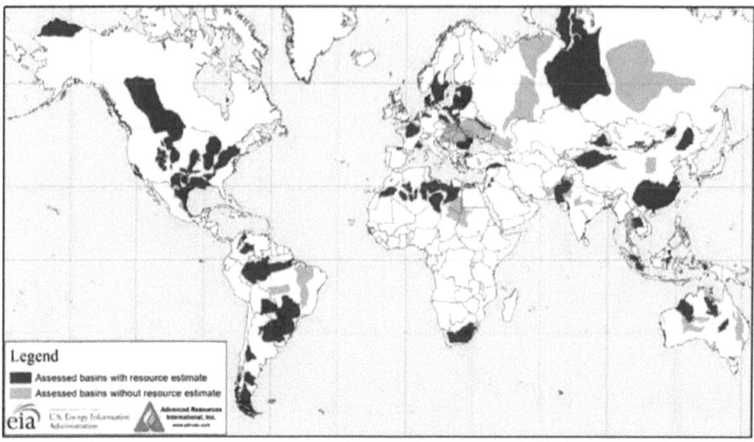

Figure 3 - World map showing sedimentary basins with potential to generate shale gas. The basins in red are those with proven potential, and those in orange are those with estimated potential. Source: EIA (2013).

However, studies and calculations of recoverable reserves in Brazil, carried out by the ANP in 2013, indicate promising figures for the country, with estimates of 414 Tcf in just five of Brazil's main sedimentary basins: Paraná, São Francisco, Parecis, Recôncavo and Parnaíba, making unconventional gas potentially mark a new moment for the Brazilian oil industry, as well as encouraging the creation of a large domestic natural gas production industry and market and a strong boost to Brazil's development in the energy sector.

In this Brazilian context of energy resources, the south and southeast regions of the country stand out for representing the largest national consumer market. One of the main fossil fuel reserves in this region is attributed to the Paraná Basin, which consists of a large sedimentary region that includes territorial portions of southern Brazil, eastern Paraguay, northeastern Argentina and northern Uruguay, totalling an area of approximately 1.5 million square kilometres. The most important geological formations in terms of oil potential in the basin are the so-called Irati Formation, which dates from the Permian period, and the Ponta Grossa Formation, from the Devonian; both lithologically made up of shales, siltstones and carbonates (lithology favourable to hydrocarbon generation). Studies carried out by the Brazilian Geological Service (CPRM, 2013) and the ANP (2013) point to a high potential for unconventional natural gas in the Paraná Basin, estimated at 226 Tcf, which would be associated with these geological formations. This potential is attributed to the accumulation of high rates of organic matter associated with a reducing environment and pressure, temperature and burial conditions suitable for generating gas in an unconventional geological model *(shale gas)* which is characterised by the low permeability of the rocks in the formation.

In the current scenario of growing demand for energy resources and electricity generation, AES Brasil, a member of the US corporation *Applied Energy Services (*AES), in partnership with the Gas Company of the State of Rio Grande do Sul (Sulgás), developed the AES Uruguaiana project with the aim of serving the consumer market in the southern region of the country (Figure 4). AES Uruguaiana is a natural gas-fired thermoelectric plant that began operating in 2000 in the city of Uruguaiana, in the south-west of the state of Rio Grande do Sul and close to the border with Argentina, and which would be responsible for supplying gas through a contractual partnership between the national oil company YPF *(Yacimientos Petrolíferos Fiscales* - YPF), AES and Sulgás. Its installed capacity is 639.9MW, but the plant has been paralysed since 2009 due to the interruption of gas supplies, justified by the Argentine government through insufficient gas pipeline capacity and infrastructure in association with the country's financial crisis. Due to the recent national water crisis, the Uruguaiana plant resumed operations on an emergency and temporary basis in 2013, 2014 and 2015, based on imports of liquefied natural gas from Central America and Africa (AES Eletropaulo, 2015). According to data provided by the company itself, the Uruguaiana thermal power plant has only been in operation for seven months since 2008, due to the lack of gas input and consequent high costs. Given this scenario, investment in and production of domestic natural gas from unconventional energy resources is strategic for the security of the Brazilian electricity matrix and for national energy independence.

Figure 4 - Image of the Uruguaiana thermoelectric plant, south-west of the state of Rio Grande do Sul. Source: AES Brasil (2015).

This master's thesis project aims to carry out further studies on the subject of unconventional hydrocarbons, covering the geology of the Paraná Basin, based on its high potential for the occurrence of this resource, and evaluating the potential of the Irati and Ponta Grossa Formations (black shales) with a view to the feasibility of supplying a thermoelectric plant with unconventional natural gas. A possible proof of the unconventional oil system potential of this formation, together with an analysis of the feasibility of this supply, could represent a new scenario in the face of the lack of natural gas input for the south and southeast of the country, as well as prospects for greater participation and production of national natural gas in the Brazilian energy sector.

CHAPTER 2

PRESENTING THE PROBLEM

The project aims to contribute to the geological assessment of formations that potentially generate unconventional hydrocarbons and reservoirs in a scenario where information about this potential is scarce and there is a shortage of natural gas supply in the national market. In this context, the study focuses on the following problem:

- What is the real potential for unconventional natural gas *(shale* gas*)* in the state of Rio Grande do Sul in terms of geological occurrence?
- What are the possible impacts related to the exploitation of this resource in the region, according to socio-economic and environmental aspects?
- What are the prospects for exploiting the resource from an economic and regulatory perspective?
- What is the average consumption of natural gas at the Uruguaiana thermal power plant to generate electricity, based on its generating capacity; and what is the potential for generating unconventional natural gas in the local geological formations (Fm. Irati and Ponta Grossa)?
- What is the feasibility of supplying the thermoelectric plant with this resource?

CHAPTER 3

OBJECTIVES

3.1. General Objective

The main objective of the research project is to assess the potential for unconventional natural gas from the sedimentary rocks of the Paraná Basin, more specifically from the Irati and Ponta Grossa formations, in terms of geological occurrence, with a view to local supply for the Uruguaiana thermoelectric plant in the state of Rio Grande do Sul.

3.2. Specific objectives

- Development of a descriptive geological model for unconventional natural gas based on the geological aspects of the main exploration fields in the United States: Barnett, Marcellus and Eagle Ford, from which a comparative analysis will be developed between these and the Irati and Ponta Grossa geological formations in the Paraná Basin, with the aim of estimating their exploration potential;
- Description and presentation of the geological aspects of the Paraná Basin with the aim of evaluating its regional geological context and the geological feasibility for the occurrence of unconventional natural gas compared to sedimentary basins proven to generate it in the United States;
- Geological characterisation of the Irati and Ponta Grossa formations, with emphasis on their organic geochemical, petrophysical and faciological characteristics, and the identification of comparative parameters with generating formations in the USA that will help to determine their potential for generating unconventional natural gas in the light of the tectono-magmatic and structural evolution of the basin, and to determine their geological viability for the occurrence of this resource;

CHAPTER 4

JUSTIFICATION AND MOTIVATION

According to Yergin (2011), natural gas can be considered a relatively safe, versatile and economical energy source. The combustion of this resource is comparatively cleaner and more efficient, resulting in reduced emissions of pollutants compared to other fossil fuels (oil and coal). The use of natural gas in the oil and energy industry has provided significant benefits for the environment, as well as reducing operating costs with machine maintenance, transport and fuel storage. It also optimises the use of raw materials and consequently improves production processes; it increases safety, equipment efficiency and the quality of the end product. Natural gas plays a substantial role in the automotive, food, textile and chemical industries, in the production of aluminium, glass, ceramics, paper and rubber, as well as in steel and metallurgical plants. It can be used in heating and cooling processes, electricity generation and cogeneration of thermal and electrical energy, and natural gas has become the main energy source in the United States over the last decade (Yergin, 2011).

According to Companhia de Gás de São Paulo (Comgás, 2013), discussions about trends and new production technologies and the use of fuels that contribute to preserving the environment have also gained momentum among government bodies. These bodies are committed to increasing the use of renewable energies and fuel efficiency, promoting cogeneration and increasing the use of natural gas due to its environmental advantages. In this scenario, the importance of policies to encourage the use of natural gas on a larger scale is growing. Its technological, safety and, above all, environmental advantages make natural gas an important alternative for improving air quality and reducing greenhouse gas emissions. This has made it easier for industries to obtain environmental quality certificates.

In this context, Brazil stands out for having several sedimentary basins with gas potential; especially with the potential for shale gas, which is equivalent to high concentrations of natural gas widely distributed throughout the country. On the other hand, as the Brazilian energy matrix is based on hydroelectric generation, there is vulnerability to climatic conditions, as in the case of the recent water crisis in the state of São Paulo, due to climate change and seasonal reductions in water flow associated with management problems, resulting in a water and electricity supply crisis for the region. In the midst of this current energy crisis, thermal power plants have been activated on an emergency basis to meet local demand for electricity. One of these plants, the aforementioned AES Uruguaiana thermoelectric plant, which is influenced by Argentinian foreign policy, has obsolete technology and operates at high costs, is the subject of this study.

The Uruguaiana thermal power station, owned by the US company AES, is supplied by natural gas from Argentina. The supply is intermediated by the YPF oil company, which was nationalised in 2012 by the Argentinian government. As a result of problems in complying with the agreement that was previously in place to supply the unit, Brazil began transporting natural gas to the region: Petrobras imports natural gas from Trinidad and Tobago, in the Caribbean region, and transports the cargo by ship over 7,700 kilometres to the regasification terminals in Argentina (A Folha de São Paulo, 2014). Sulgás, (Gas Company of the State of Rio Grande do Sul), which is responsible for supplying natural gas to the region, buys the resource from Petrobras and injects it into the neighbouring country's pipeline network, considerably increasing production costs.

Given this scenario, assessing the geological potential for shale gas in the Paraná Basin is a strategic plan to guarantee energy supply for the south and southeast regions of Brazil, which account for the largest demand for electricity in the country. In addition, studies on the Brazilian potential for exploration, production and distribution of this resource play an important role in Brazilian energy security, as well as encouraging the industry and the national market for natural gas, a potentially more efficient and less polluting resource, fully justifying the development of this project.

CHAPTER 5

WORK PLAN AND METHODS USED

The work plan consists of the following steps:

- Bibliographical survey on the regional geological aspects of the Paraná Basin;
- Literature survey on unconventional natural gas, with a focus on shale gas from its historical context, current scenario, geological aspects and associated impacts;
- Data collection on the main shale gas producing fields in the United States (Barnett, Marcellus and Eagle Ford), considering geological aspects and developing a descriptive and comparative model about them;
- Survey of the comparative parameters established by the USGS to analyse the geological viability and occurrence of this resource;
- Survey of these parameters in the black shales of the Paraná Basin (Irati Formation and Ponta Grossa Formation) and in the shales of the American producing fields for comparative purposes;
- Comparative analysis of these parameters, summarisation of this study in a table (descriptive model for the geological feasibility of shale gas) and specification of this analysis for the state of Rio Grande do Sul due to its proximity to the case study (thermal power plant in Uruguaiana);
- Quantification of the average consumption of natural gas by the AES Uruguaiana thermoelectric plant as a result of its electricity generating capacity and the potential production of unconventional natural gas from these geological formations in order to analyse the viability of this supply;

CHAPTER 6

OIL SYSTEMS

6.1. Concepts and Definitions

Over decades of exploration, the oil industry has realised that the occurrence of hydrocarbon deposits of significant volume is associated with certain geological requirements that occur simultaneously in sedimentary basins. The study of these characteristics in an integrated manner and the preliminary simulation of the optimum conditions for their existence concomitantly with the reduction of the exploration risks involved in drilling wells, which are high-cost operations, were consolidated into a single concept: the petroleum system (Magoon & Dow, 1994). According to the proposal by Magoon and Dow (1994), the concept of a petroleum system has been defined as a natural system that includes all the geological elements and processes necessary for the accumulation of hydrocarbons. These geological elements correspond to the set: source rock, reservoir rock and sealing rock, which must coexist and be synchronous with the processes of generation, migration and accumulation of oil, associated with the formation of traps (Figure 5).

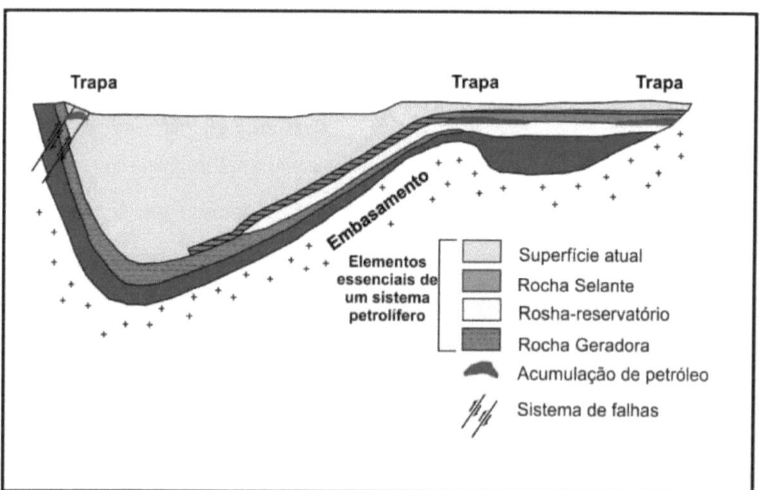

Figure 5 - Schematic representation of a petroleum system and the elements that condition its formation. Source: Magoon & Down (1994) apud Silva

(2013).

According to Chapman (1976), petroleum geology broadly corresponds to the geology of sedimentary basins, where the main accumulations of hydrocarbons occur. Knowledge of sedimentary basins is therefore essential for understanding the petroleum system and the processes that act on it.

Petroleum geology is not only based on geological studies, but also on experiments and the practice of the oil industry. While geology studies processes such as the generation, migration and accumulation of hydrocarbons, in their respective order, the industry studies them in reverse order, focussing its studies on tracing, which is not just interpretative like the other processes, but observable, concretely defined and measurable. Geological studies show that oil tends to accumulate in deeper sedimentary packages (sedimentary basins) than in areas of thin sedimentary packages, and that it tends to be found in rocks dating from the Mesozoic to Triassic geological periods, and is more common in anticlinal traps than in any other type of traps (Chapman, 1976). However, experience shows that the accumulation of hydrocarbons does not follow a fixed rule, and it is possible to find oil fields in scenarios that do not follow traditional traps, such as the case of unconventional hydrocarbons studied here, where the reduced permeability of source rocks and reservoirs predominantly plays the role of a trap.

This dissertation will first describe the conventional model of hydrocarbon accumulation, following a traditional model of petroleum systems. Next, an unconventional model will be presented, exemplified by the accumulation of natural gas in low-permeability reservoirs, such as shale gas, the focus of this study.

6.2. Geological elements

6.2.1. GENERATOR ROCK

According to Milani et al. (2001), the fundamental element for the occurrence of oil is the existence of considerable volumes of suitable organic matter simultaneously accumulated in the deposition and formation of certain sedimentary rocks, known as source rocks. It is these rocks that, when subjected to specific pressure and temperature conditions, are responsible for generating hydrocarbons in the subsurface. Source rocks are considered to be the main element for the occurrence of hydrocarbons, because when they are absent in a sedimentary bay, there is no possibility of substitution, unlike the other constituent elements of the petroleum system, which even when absent, can be compensated for in some way by exceptional geological conditions or by suitable coincidences, as in the case of unconventional petroleum systems, better detailed later in the dissertation.

Source rocks are predominantly made up of very fine-grained detrital material (clay fraction), and are generally represented by shales and/or carbonates; indicators of low-energy sedimentary palaeoenvironments that, for various reasons, experienced explosions of microscopic life, which are a source of organic matter. Autochthonous (planktonic material) or allochthonous (terrestrial plant material transported into the environment) organic remains are incorporated into the unconsolidated sedimentary material in the form of diluted organic matter. In principle, the greater the amount of organic matter, the greater the rock's potential for generating hydrocarbons. However, the incorporation of this organic matter into the rock must come from

For this to happen, the environment must be very depleted or free of oxygen, an oxidising element that destroys the concentrations of C and H in the original organic particles. In general, anoxic palaeoenvironments favour the preservation of organic matter and, consequently, the generation of oil. Generally, common sedimentary rocks have Total Organic Carbon (TOC) levels of less than 1%. According to Silva (2007), the minimum TOC values for a rock to be considered potentially generative are approximately 0.5 to 1.0%, throughout the entire geological unit. The type of oil generated depends fundamentally on the type of organic matter preserved in the source rock. Organic matter derived from higher plants tends to generate gas, while material derived from zooplankton and phytoplankton, whether marine or lacustrine, tends to generate oil (Milani et.al.2001).

Organic matter content is the main requirement for hydrocarbon generation in sedimentary basins. Generally, source rocks contain between 0.5% and 2.0% of their weight in organic carbon for shales, and up to 0.2% for carbonates; source rocks considered to have good oil potential contain TOC values of around 12% (Milani et al. 2001). According to Milani (op.cit.), the main conditions for sediments to become source rocks are:

6.1.1.1. The presence of adequate and sufficient organic matter (TOC%).

6.1.1.2. The right type of kerogen, whether it has oil-prone or gas-prone potential.

6.1.1.3. Level of maturation necessary for the formation of oil.

6.1.1.4. Organic Matter

The main representatives of organic matter favourable for hydrocarbon generation are plankton (marine and fluvial) and bacteria, which also account for the majority of organic matter preserved in sediments. The organic matter suitable for generating oil is represented by marine phytoplankton, due to its composition rich in lipids (fats). The first stage in the incorporation of organic matter into sediments is the generation of this organic matter, with photosynthesis being the essential biochemical process for producing the precursors of sedimentary organic matter that will result in potential hydrocarbon source rocks. Once this first condition is met, the main factor becomes the existence of an anoxic environment (Milani et al. 2001).

Approximately 98% of the organic matter present in sediments corresponds to an amorphous, insoluble material called kerogen. The term kerogen was first used in 1912 by Crum & Brown (apud Link, 1996) to describe pyrobituminous shales in Scotland. Nowadays the term is used for the fraction of sedimentary organic matter that is insoluble in the presence of organic solvents. According to Link (op.cit.), kerogen is an insoluble, high molecular weight polymer compound that comprises around 90% of the organic matter in a shale. The remaining 10 per cent is bitumen of varying compositions which, according to some authors, corresponds to thermally altered kerogen. Other authors maintain that, even under the most favourable conditions, a maximum of 15% of sedimentary organic matter will be converted into liquid hydrocarbons. Of this 15%, no more than 15% of the hydrocarbons will migrate from the source rock, i.e. around 2.2% of the total. The amount of oil generated that will form commercial accumulations also corresponds to this percentage range.

6.1.1.5. Ripening

The stage of thermal maturation of a source rock, i.e. the temperature it must be subjected to in order to generate oil, will also influence the type of oil generated. Under normal conditions, a source rock begins to transform kerogen into oil at around 60°C. Initially, a low-maturity, viscous oil is formed. As the temperature rises, the oil generated becomes more fluid and the amount of gas increases. At around 90°C, the source rocks reach their generation peak, expelling large quantities of oil and gas. As the temperature rises to 120°C, the oil becomes increasingly fluid and richer in dissolved gas. In this temperature range, the amount of gas predominates over the oil generated, and the product is considered a condensate. Between 120°C and 150°C, only gas is generated by the source rocks (Link 1996).

6.1.2. RESERVOIR ROCK

Reservoir rocks are generally of detrital composition and sand to pebble grain size, characterising a porosity in which oil can be stored and later extracted, and are representative of high-energy sedimentary palaeoenvironments. These rocks usually correspond to sandstones, calcarenites and various conglomerates. However, any rock that contains adequate porosity, not necessarily intergranular, but of a different nature, such as from fracturing or dissolution, can also play the role of reservoir rocks (such as fractured crystalline rocks, leached marbles, among other examples). The most common porosity values among reservoir rocks range from 5% to 35%, with the latter concentrated in the 15% to 30% range (Link,

1996). The porosity of these rocks not only serves to store oil, but also as important oil migration routes. In order to be efficient, an oil reservoir must be relatively porous and permeable, as well as containing some kind of barrier (trap) to contain the migration of the fluid. The process of compacting sediments results in a reduction in the rock's porosity and permeability. However, when sediments are saturated with non-compressible fluids (such as water, oil or natural gas), porosity will only be reduced when part of the fluid is released. Therefore, the rate of compaction is directly related to the compressibility of the fluid and the rate at which it is expelled, which in turn is controlled by the permeability or hydraulic conductivity of the rock being compacted (Chapman, 1976).

Compaction and hydraulic conductivity are factors related to each other and to the porosity and permeability properties of rocks. They are also important for understanding oil reservoirs, especially in the cases of unconventional accumulations presented in this work, represented by low-permeability reservoirs.

6.1.2.1. Porosity

Sedimentary rocks are made up of grains of varying sizes that are grouped together by physical or chemical properties (cementation). The empty spaces between these grains correspond to the rock's porosity, measured as a percentage of the rock's total volume. The maximum volume of fluids that a reservoir rock can contain is equivalent to the volume of pores present in the rock, thus realising the importance of determining porosity accurately. According to Khamsin (2004), due to the variety and complexity of sedimentation processes, different types of porosity can be generated in a rock. These are:

Effective porosity: When the empty spaces are connected and have flow potential through them.

Non-effective porosity: When the porosity is isolated (unconnected). Primary porosity: Also called original porosity. This is the porous space that the rocks have at the time of deposition. It is usually a function of the amount of space between the rock-forming grains. Original porosity is reduced by compaction and diagenetic processes related to groundwater.

Secondary porosity: Caused by solutions (groundwater), recrystallisation and fracturing. Secondary porosity develops after sedimentary deposition.

The porosity of a rock is directly related to the compaction exerted on it. Compaction is part of the diagenetic process and begins with burial and can continue to depths of approximately 9 kilometres. Its main effect is to increase the density of the rock and reduce its porosity. In the case of sandy material, compaction results in minimal losses of porosity and permeability; due to the properties of quartz, the deformation is predominantly elastic (reversible as the weight exerted by the upper package decreases). The compaction of clay material results in a high and permanent reduction in the porosity and permeability of the rock due to the plastic deformation characteristic of clays (irreversible). The compaction of carbonate rocks occurs at different levels, according to the proportion of plastic material in their composition (they can show behaviour similar to that of clays, but generally suffer little mechanical compaction) (Khamsin, 2004).

6.1.2.2. Permeability

The permeability of a rock is the measure of how easily it allows fluids to pass through. Permeability is a porosity-dependent factor: the greater the porosity, the greater the permeability of a rock. However, permeability also depends on the connectivity between the pores, the size and shape of the grains and their distribution, as well as factors such as the capillary force, which is related to the saturation of the rock (Chapman, 1976). Based on Chapman's information (op.cit.), it can be stated that:

- The greater the porosity, the greater the permeability;
- The smaller the grain size, the lower the porosity and permeability;
- The smaller the grains, the greater the contact surface between them, resulting in greater friction between the percolating fluid and the rock and, consequently, lower permeability.

Permeability is measured in darcies (Darcy): $Q = (k/\mu)\ A\ (p1 - p_2)\ /B\ L$; where:

- Q: flow rate (bbl/d)
- K: permeability (Darcy)
- μ: fluid viscosity (centipoises)
- A: flow area in square section $(ft)^2$
- p_1 - p2: pressure difference (psi)
- L: length of flow path (ft)
- B: volume formation factor (bbl/bbl stocked)

Permeability is also influenced by capillary pressure and relative permeability:

Capillarity: When the pores of the rock decrease in size, the surface tension of the fluid in the rock increases. If there are several fluids in the rock, then each one has a different surface tension, which shows a variation in the pressure exerted between them (capillary pressure). Normally, the capillary pressure is sufficient to prevent the flow of one fluid in the presence of another.

Relative permeability: When water, oil and gas flow through permeable reservoirs, the flow rates can be altered by the presence of other fluids. A single fluid flows through the rock with its own flow rate, but the presence of another fluid can alter this flow rate. In this way, it can be said that the flow rate can be altered by the quantity and type of fluids present in the rock, i.e. the degree of fluid saturation of the rock and the nature of these fluids, which influences their ability to reduce the pore space.

As the burial rate increases, the greater weight exerted by the sedimentary package intensifies the compaction process, which rearranges the sediments in a more cohesive manner, reducing the porosity of the rock and/or isolating some pores from others. However, compaction not only reduces the volume of pores in a rock, it can also impede the flow or release of the fluid present, reducing its permeability. In this way, the fluid is trapped in isolated pores in the rock, or by the reduced porosity of the rock itself (Chapman, 1976). This is the case of so-called low-permeability reservoirs, exemplified by Basin Centred Gas or Tigh Gas (gas in low-permeability strata), or Shale Gas, trapped in the source rock itself due to its reduced permeability; these are discussed further in this paper.

6.1.2.3. Reservoir characterisation

Sedimentary rocks are the essence of oil reservoirs: they are the only rocks capable of being both generators and reservoirs, and they are also the only ones with sufficient primary porosity and permeability for this condition; and consequently they are responsible for the largest amount of oil in storage. Other rocks, such as igneous and metamorphic, can only act as reservoirs on smaller accumulation scales. The essential conditions for a rock to have reservoir potential include adequate porosity and permeability, as well as other geological factors that influence reservoir properties. These are: weathering, erosion and deposition, which determine the character of the reservoir; granulometric factors such as grain size and shape; type of cement and matrix; diagenesis (compaction rate); fracturing; fluid pressures, as well as the size and type of reservoir in question. These factors are important as they are directly related to the porosity and permeability of a reservoir (Link, 1996). According to Link (op.cit.) these factors influence reservoir quality as follows:

Grain size: In clastic rocks, grain size can be quite variable. This variation is related to factors such as mineral composition, weathering, depositional environment, as well as the distance and intensity of sedimentary transport. In this context, it can be said that larger uncemented grains usually indicate greater porosity and permeability than smaller grains. Rocks with a more uniform grain size are more porous and permeable than those with a variety of grain sizes. In general, rocks with well-selected grains are better reservoirs than rocks with less selection.

Grain shape: This property is directly related to porosity: spherical grains tend to increase porosity, while angularity closes the space between the grains, reducing permeability: when the grain orientation is unidirectional, there is an increase in permeability along the major axes (lateral permeability), and a reduction along the minor axes (vertical permeability). This variation in permeability can be exemplified in pelitic rocks: flattened grains of shale do not show normal permeability to fissility, but do show lateral movement of fluids during compaction.

Grain compaction: Spherical sandy grains in an open arrangement can generate porosity of almost 48 per cent. However, the same grains in a compacted arrangement show porosity of almost 26 per cent, which does not change even when the load on the upper layers changes. Well-sorted and well-compacted spherical grains produce greater porosity than angular and poorly-sorted grains.

Matrix: The composition of the matrix influences the internal volume of the rock: a decrease in the grain size of the matrix and an increase in the degree of penetration of the matrix reduces porosity and permeability. In fact, the fine material reduces the pore spaces between the larger grains. In addition, the materials that make up the matrix are usually altered by diagenesis, surface water or incipient metamorphism. Subsequently, porosity and permeability are reduced due to changes in minerals and internal volume.

Vaccular porosity: Carbonate rocks are sensitive to contact with solutions, both during diagenesis and post-diagenesis. The dissolution of carbonate rocks can result in the formation of caves, granular dissolution or fossil components of the original rock. This dissolution corresponds to the vacular (secondary) porosity of the rock and can improve the original reservoir potential, as it increases storage space due to the increase in porosity. However, the vacular porosity of carbonates can subsequently be lost due to the mineral infiltration that occurs with the precipitation of late solutions.

Fracturing: Rocks, both crystalline and sedimentary, can have their reservoir potential increased due to the presence of fractures. Fracturing is related to regional tectonic stress and occurs in specific patterns, which can be exploited in recovery programmes. However, the fractures must be sufficiently developed to allow the transmission of fluids through the rock. Burial pressure often causes fractures to close at depth. When this happens, the producer can use hydraulic fracturing to reopen and maintain fluid passages in communication with the existing natural fractures (a technique discussed in detail later in the dissertation). These openings allow fluids to pass through, but they can also impede them. Knowledge of fracturing is therefore essential to oil production activities.

Cementation: Intergranular cement can be composed of any mineral material capable of being distributed through a granular, non-cemented and porous medium. Cementation can correspond to a diagenetic process, and to the precipitation of mineral material in the intergranular spaces in order to hold the grains together. The amount of cement determines how much porosity and permeability have been reduced due to the compaction of uncemented material. Cementation can be partial or complete. When cementation is complete, the rock has no reservoir potential, unless solution and secondary cement dissolution or grain growth occur. On the other hand, the amount and type of cement can induce fracturing of the rock, which can increase its porosity.

Type of reservoir: Reservoirs are formed under different conditions, which result in different types of rock, grain size and shape. These differences mean that reservoirs cannot be exploited using a single set of techniques, and it is essential to study them in order to determine the general characteristics of each type of reservoir and the respective techniques best suited to exploitation. Hydrocarbon reservoirs are generally clastic or carbonate in nature. The most common clastic reservoirs consist of sandstones and conglomerates made up of quartz, feldspar and a small amount of heavy mineral grains. In general, these grains are very mature, well rounded and well sorted due to their resistance to mechanical and chemical destruction. Based on these characteristics, clastic rocks have good porosity and permeability and generally make excellent reservoirs. Carbonate rocks also have great reservoir potential, which is comparatively greater than clastic reservoirs, although their occurrence is less common.

Reservoir size: Oil recovery programmes involve appropriate well spacing and drilling intervals, which are related to knowledge of reservoir size. Geological structures that occur as reservoirs are direct functions of their depositional environments and the length of time that the appropriate conditions of sedimentation, transport, etc. have persisted. Knowledge of the relationship between geological factors and reservoir structure generates dimensional information that can be used for the proper spacing of wells and the stabilisation of zones that will be drilled in order to achieve maximum oil and gas production.

Fluid pressure: The loading of the upper layers (burial pressure), gas in solution and meteoric water are the reasons for the development of pressure in reservoirs. When a reservoir is in production, the internal pressure of the system is reduced and the fluid phases are changed, also altering the initial production rates. These production rates can be altered by various factors involving both the fluids and the reservoirs. As reservoir pressure is reduced, the gas in solution separates from the liquid phase (oil). The pressure at which this phenomenon occurs is called the gasification point. The ratio between the volume of gas produced and the volume of oil produced is called the gas-oil ratio. As the reservoir pressure decreases, the gas-oil ratio increases until all the gas is used up. Reservoir pressures are initially sufficient to force the fluids to the surface. When the reservoir pressure drops due to production, the gas dissolved in the oil is released forming a gaseous phase. As expansion takes place, the fluids remaining in the reservoir are rapidly driven towards the surface concomitant with the rapid decline in pressure.

6.1.3. SEALING ROCK

Once oil fluids have migrated into a reservoir, they must find conditions that prevent the migration process from continuing. Normally, this condition is provided by low-permeability rock layers located above reservoir rocks, trapping the oil and resulting in an oil accumulation: these rocks are called sealants, and are fine-grained (shales, siltstones, calcilutites) or any rock characterised by low permeability, whose fluid transmissibility is lower than that of the reservoirs. Occasionally, facies or diagenetic changes within the reservoir rock itself, or even structural elements such as faults, or differences between fluid pressures, can also serve as a seal for oil (Link, 1996).

6.3. Processes

6.3.1. GENERATION

Oil is generated from the breakdown of kerogen molecules in the source rock as a result of the increase in temperature over a certain period of time. High temperatures (ideally between 110 and 150 °C for oil, or up to 180 °C for gas) induce the maturation and expulsion of oil, which then undergoes upward movement by thrust, when under hydrostatic conditions, or by flotation associated with fluid flow under hydrodynamic conditions, representing the primary migration stage (Link, 1996).

According to Rostirolla (1999), when saturation in residual oil reaches 0.5% to 4%, the forces of thrust and internal pressure condition the movement of hydrocarbons. Saturation is always higher at the top than at the bottom of the layer and depends on the dip angle, viscosity, hydrocarbon supply rate and permeability. The absence of generation is a restrictive condition for the functioning of the petroleum system. However, even if a rock contains organic matter, this does not necessarily represent a variable indicative of the generation factor. For this to be the case, it must fulfil the appropriate conditions for generation for the oil system model considered, and the presence of three requirements is essential: quantity of organic matter, quality of the organic matter and maturation. The expulsion of hydrocarbons from the source rock occurs when the saturation in shales (fraction of the microporosity occupied by hydrocarbons) exceeds 10-20%. Normally, expulsion efficiency increases with increasing maturity, reaching 80 to 90 per cent for rich, mature rocks. In fluid overpressure processes, oil can be expelled by hydraulic fracturing, as in the case of unconventional hydrocarbon exploration.

The main criteria used to characterise the variables indicative of generation are mainly based on the total organic carbon (TOC) content, or relative measure of the carbon present in the form of kerogen and bitumen, as well as the combination and relationship between the hydrocarbon content already present in the rock, hydrocarbons generated in the rock (from the breakdown of kerogen at high temperatures), and volatile components (CO2 and H2O). In addition to the quantity and quality of organic matter, Rostirolla (1999) also considers the volumetric capacity of the source rock.

6.3.2. MIGRATION

Once the oil is generated, it occupies a larger space/volume than the original kerogen in the source rock. The latter becomes supersaturated with hydrocarbons and the excessive pressure causes the source rock to fracture intensely, allowing the fluids to be expelled to lower pressure zones. The flow of petroleum fluids through different routes through the subsurface until they arrive at a place where there is porous space, sealed and trapped, suitable for storing them, is the phenomenon of migration. The usual routes in a sedimentary basin are fractures of varying scales, faults and different porous rocks, linking the sources of generation, which are deep and under high pressure, to shallower fluid focussing regions with lower pressures (Rostirolla, 1999).

Migration corresponds to the movement of oil, which occurs when the fluid pressure exceeds the capillary pressure, which restricts flow. Migration routes behave like a conduit, creating different pressure gradients and allowing the fluid to move. According to Rostirolla (1999), the tendency for migration to occur from faults and fractures is lower than along layers. The author states that tectonic permeability is less effective than sedimentary permeability. However, the main factor influencing migration is the pressure gradient, as well as the ease of flow convergence, factors that can generate atypical situations for migration. The concept of migration is scientifically accepted, however, there is not total agreement between authors as to the mechanisms involved and the distances travelled. Based on conventions and definitions by Link (1996), it can be concluded that there are three different types of migration:

Primary Migration: When the oil moves from the source rock to the reservoir rock.
Secondary Migration: When the oil moves within the reservoir rock after its accumulation.
Tertiary Migration: This occurs when the oil is not stored and continues to move in search of lower pressure zones. The oil ends up as ooze, oxidised or suffering bacterial degradation.

6.3.3. ACCUMULATION

Once in motion, petroleum fluids are directed towards areas of lower pressure, which are usually positioned structurally higher than their surroundings. The geometric configurations of sedimentary rock structures that allow migrating fluids in the surrounding area to be focussed in high places that do not allow them to escape in the future, forcing them to accumulate there, are called traps (Link, 1996).

They can be as simple as the flank of homoclinals or, more commonly, as the apex of anticlinal folds/arcs/salt domes, or even complex situations such as the superimposition of folds and faults of various kinds. This type of trapping in an elevated structure is called structural trapping. However, oil is not always trapped under structural conditions. Occasionally, the migration of oil can be stopped by the wedging of the carrier layer or its blockage by a diagenetic barrier or a barrier of different permeability, so that it is trapped in structurally unremarkable positions. In this case, we have stratigraphic trapezing (Milani, 2001).

Understanding the characteristics of the deposition and integrity of the sealing rock is fundamental to understanding the favourable locations for oil accumulation. The presence (or absence) of the seal as an indicative variable requires the adequate existence of the following physical properties: capillarity (minimum pressure that supports the buoyancy force of the fluid), lateral continuity (geometry and facies architecture), lithology (occurrence of halite, anhydrite, shales, carbonates or chert), ductility (tendency of the rock to flow and respond plastically during deformation), thickness (maintenance of depositional characteristics over time and sedimentation rate). The trap is therefore a geological entity that integrates a reservoir rock with shape, adequate volumetric capacity and an effective seal. The volume of oil and gas preserved depends on predicting the size and type of the trap which, together with migration, is the factor with the greatest influence on geological success. Traps can therefore be categorised as structural, stratigraphic and hydrodynamic, with stratigraphic traps being where most of the resources are found (Link, 1996).

Over the last few decades, oil exploration and production has undergone significant advances, not only in terms of technology, such as drilling technology, but also in terms of the concept of petroleum systems: currently, due to economic factors and the relationship between supply and growing demand for energy resources, different concepts of petroleum systems have been implemented and analysed in terms of their potential for hydrocarbons. In this context, oil is now being explored from sequences that were previously not significant or economically viable for exploitation. These are the so-called unconventional petroleum systems, which are establishing a new exploration frontier in Brazil and around the world.

6.4. Conventional Petroleum Systems

The concept of conventional petroleum systems corresponds to the definition of petroleum systems proposed by Magoon & Dow (1994), which includes the presence of source rock, reservoir and sealant elements, concomitant and synchronous to the occurrence of hydrocarbon generation, migration and accumulation processes. In these so-called conventional systems, the accumulated hydrocarbons can be produced using available technology, without the need for specific recovery mechanisms (the oil produced is extracted by the well using a pressure difference). In general, primary and secondary recoveries are used for conventional production and recovery rates vary between 30 and 40 per cent. Conventional reservoirs are characterised by their good porosity and permeability, low viscosity and identifiable water interface. They are mostly sandstones and calcarenites (das Virgens, 2011).

6.5. Unconventional Petroleum Systems

Unconventional petroleum systems are those in which normal production flows are not considered economically viable (Figure 6). This includes reservoirs where there is a need for massive stimulation, since production from these reservoirs requires special recovery techniques, such as steam injection or hydraulic billing technologies; due to their low recovery rate (less than 20% according to conventional exploration techniques). Unconventional reservoirs are exemplified by: tight gas or basin centred gas (BCG), *coal bed* methane *(CBM),* heavy oils and *shale* gas (Yergin, 2011).

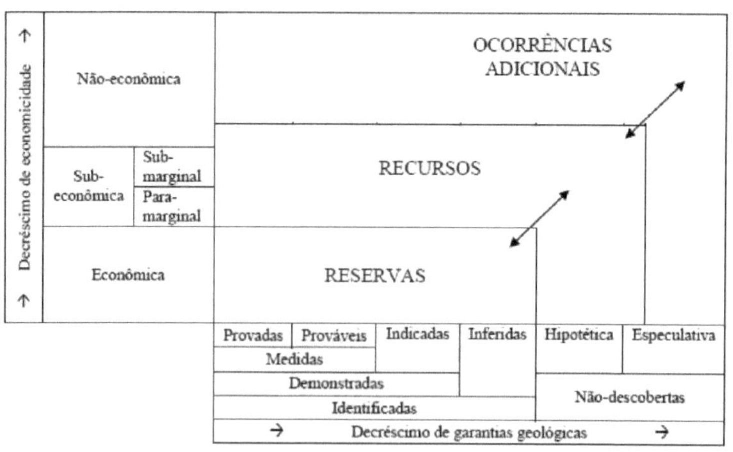

Figure 6 - Classification of mineral resources according to geological and economic factors. Source: Malagueta (2009).

When the volume of reservoirs is partially or totally estimated and there is extraction technology, these resources can be called Technical Reserves or Technically Recoverable Reserves, which means that the existence of the resource is recognised and that there is extraction technology. However, until the reserves are developed, wells are drilled and production takes place, the reserves are not classified as proven, probable or possible (das Virgens, 2011).

It is possible that due to the low permeability of the source rock or the formation as a whole, the products generated during catagenesis (oil and/or natural gas) do not enter into secondary migration and there is, in fact, partial primary migration. If this happens, the following situation is possible: the source rock has low permeability to the point of drastically reducing primary migration (< 0.1 mD) (Table 1). It is possible that in this situation oil in shale (first) or gas in shale (second) has already been generated, natural gas being the gaseous phase of oil. The generator then also becomes the reservoir and migration is restricted to the same rock. Here, all the elements of the petroleum system are present on a reduced scale, but migration has not taken place from the generator to the reservoir. This type of petroleum system is classified as unconventional (Malagueta, 2009).

Table 1 - Comparison (in mD) between the permeabilities of conventional and unconventional reservoirs (gas in shale and coalbed methane - CBM).

Comparison of Permeabilities (mD)	
Gas in shale	<0,1
CBM	**>0,1 <50**
Reservoir rocks (siliciclastic and carbonate)	**>100 <10.000**

Source: Modified from Malagueta (2009).

However, it is more common for oil to be generated, migrated and accumulated, but in a geological formation that differs from its neighbours due to lateral and vertical regional lithological barriers, including migration routes. The same basin can have several such petroleum systems, which are known as Independent Petroleum Systems (IPS) (Magoon & Dow, 1994) or unconventional petroleum systems. Magoon & Dow (op.cit) state that for each basin explored there can be around three or four IPSs. However, as mentioned above, hydrocarbon accumulations in shale present the elements of a petroleum system, but on a reduced scale, especially with regard to migration (primary migration). From this perspective, the application of Magoon & Dow's (1994) concept of independent petroleum systems becomes valid or coincidental for these oil accumulations in shales, classifying them as unconventional or independent petroleum systems.

CHAPTER 7

UNCONVENTIONAL OIL

7.1. History

Since the beginning of the 21st century, a fear has threatened the world's energy sector and, consequently, global economic stability: the fear of the end of oil. Oil is a finite and non-renewable fuel, if we consider the time scale of humanity, and has been responsible for financing the significant economic and population growth of the century. The fear that oil reserves were coming to an end arose with the peak oil theory, proposed by M. King Hubbert[2] in 1956 and based on the idea that the peak of oil exploration had already been reached. The reserves were therefore running out as production reached its peak. The theory represents world oil production graphically by a peak (or bell), which corresponds to maximum exploitation due to growing demand, a phase that is either nowadays or about to begin, followed by an inevitable decline in production due to the depletion of the resource. However, the peak oil theory does not take into account the prospects for technological advances and scientific knowledge, which are decisive factors in this scenario because they are capable of guaranteeing decades of production until a decline begins, which would be much less abrupt under these circumstances. Innovations in the oil sector have generated controversy in the peak oil model, and indicate the representation of production as a plateau, i.e. the peak of exploration can be extended as shown in Figure 7, postponing the decline due

[2] American geologist associated with the *Shell Oil Company* during the 1950s. Throughout his work experience, he concluded that the discoveries of oil reserves, when represented graphically, were bell-shaped, as were the curves relating to the production of this resource. Based on these observations, he established the theory of *peak oil,* stating that the oil sector would reach a peak in exploration and production followed by an abrupt decline in productivity due to the depletion of this resource. In 1962, Hubbert predicted that the production of 48 American states would peak between 1965 and 1970, information which, despite its low credibility, would have materialised. However, this was later justified by the scarcity of technology and information about new reserves rather than the exhaustion of the fossil resource. Modern adaptations of Hubbert's theory suggest the shape of a plateau to represent this scenario, while other researchers claim that the peak represents the end of economically accessible production and can be applied to different moments in society, such as the case of unconventional oil resources.

to depletion of reserves, which would be much more gradual and less impactful on the market (Yergin, 2011).

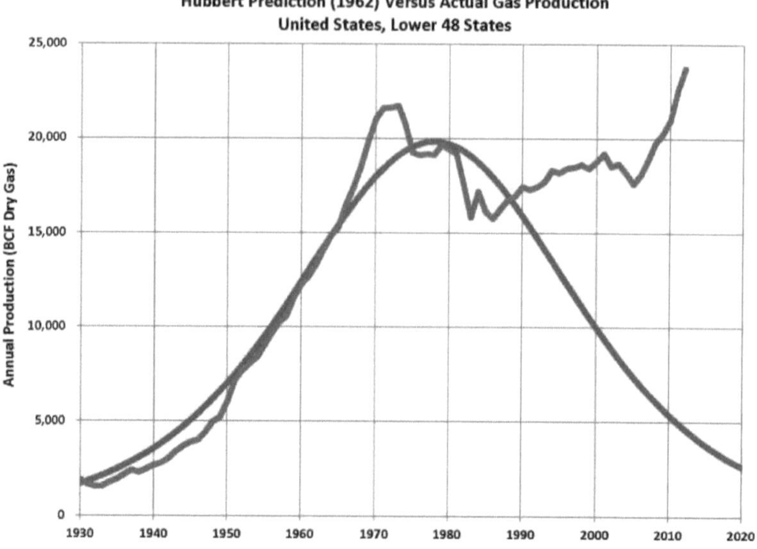

Figure 7 - Schematic representation of Hubbert's 1962 forecast for oil and natural gas production in 48 American states (in red), versus production in 2012 (in blue). Source: Modified from Hubbert, (1962), apud "Energy Resources," National Academy of Sciences, Publication 1000-D, p.81-83.

According to Yergin (2011), although it may seem recent, the fear of the end of oil has been present throughout world history, always associated with the growing demand for this resource. Started by Edwin Drake in northern Pennsylvania, the modern oil industry has faced challenges in balancing supply and demand for decades. Initially used as a source of light to replace whale oil, petrol became essential to society in 1859. Since then, demand for the fuel has risen sharply with the development of the automobile and war industries, among others, in order to meet the rapid growth of emerging economies (Figure 8). In this context, the oil industry has gone through crises, resulting in price rises and threatening world economic stability at various times in history, crises that would have been alleviated or solved by research and technological advances to guarantee production. The current situation is no different: is oil really coming to an end, or is there a lack of technology to continue exploiting it?

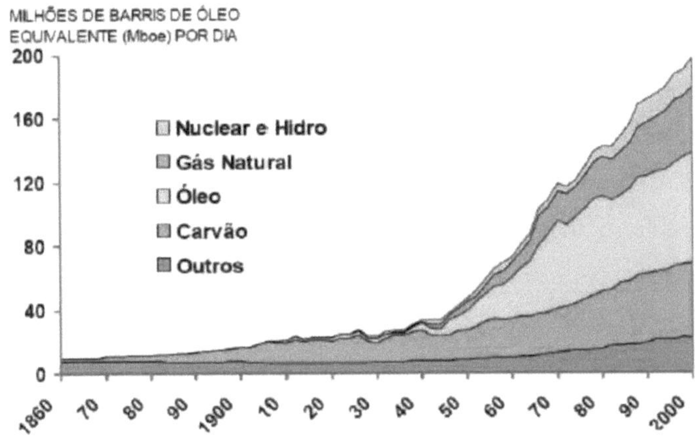

Figure 8 - Representative graph of world primary energy consumption from 1860 to 2000. Source: IEA (2007) apud. Malagueta (2009).

The history of the oil and gas industries is strongly linked to technological advances. New technologies have been developed in order to identify new reserves and also to produce from fields that have already been explored. It is known that less than 60 per cent of the accumulation in oil fields can generally be produced using traditional methods (Yergin, 2011). Faced with this situation, research has been carried out to increase this recovery rate, based not only on exploration technologies, but also on studies of unconventional oil sources.

7.2. Current Scenario

There are different unconventional sources of oil and natural gas. Even though the occurrence of these types of accumulations has been recognised for decades, the challenge lies in finding techniques to recover this resource on a commercial scale.

Studies on petroleum geology have been divided into two branches of hydrocarbon research: conventional and unconventional. These represent different processes for accumulating and characterising reservoirs, as discussed earlier in the chapter; as well as corresponding to the new prominence in the theory and concepts of petroleum geology. The geological study of conventional hydrocarbons focuses on identifying the relationship between source, reservoir and sealant rocks and their association with the processes of migration, trapping, accumulation and contemporaneity. The geological study of unconventional hydrocarbons focuses on evaluating the continuity of the oil accumulation process, associated with lithological and structural analysis and the physical properties involved in the process (Zou, et.al. 2013).

According to the International *Energy Agency* (IEA, 2012), the concept of unconventional hydrocarbons includes: *gas and oil in shale, tar sands, synthetic oils and derived products, gas and oil generated from mineral coal and biomass*, as well as *hydrocarbons at different levels of maturity such as kerogen, heavy oils and condensates.*

New in the natural gas sector is the so-called *Shale Gas:* unconventional natural gas generated and trapped in shales due to their reduced permeability. This type of accumulation has

The oil and gas industry has stood out among unconventional energy sources due to its potential for high concentrations of natural gas and its wide geographical distribution. Exploration, which was previously economically unviable, has become possible due to technological development: techniques such as hydraulic billing and horizontal drilling have made it possible to exploit this resource and have brought significant economic returns.

Along with other unconventional sources of oil and gas, shale gas has represented a new exploratory frontier, different from traditional conventional production, but one that could have a significant impact on the world economy. According to Yergin (2011), this new definition of oil sources has been expanding and helping to meet society's growing demand for energy resources. Yergin (op.cit.) believes that by 2030, these unconventional sources will be the basis of North American oil production and will be renamed conventional.

7.3. North American experience

The constant search for energy resources has resulted in several advances in the unconventional hydrocarbons sector. Research carried out by the U.S. Geological Survey found that natural gas could not only be exploited in conventionally configured reservoirs, such as anticlinal traps, but could also be found in low-permeability rocks. A property that gives the same rock package both the role of reservoir and sealant. The discovery revolutionised the oil industry in the United States, but it still had to face several difficulties in its exploration process: there was no knowledge of the techniques needed to exploit this resource, resulting in high costs for research and the implementation of new technologies (Yergin, 2011).

The main exploration techniques for unconventional systems were developed in the United States based on production studies carried out in the Bakken Shale[3] and are known as fracking and horizontal drilling. The term fracking refers to the hydraulic fracturing technique, which has been used since the 1940s and basically consists of injecting a large volume of water under high pressure, together with sands and chemical compounds, into a geological formation saturated with oil. Its aim is to fracture the rock in question in order to create a passage route for the explored fluid. Another exploration technique, called horizontal drilling, emerged in the 1980s. Technological advances allowed operators to drill to certain depths and then drill at an angle, or even completely horizontally. This technique allowed greater access to the reservoir, resulting in greater recovery of gas (or oil). Based on this knowledge, companies began to combine both drilling techniques, bringing innovation to the market and great financial returns. However, this financial return can be associated with high environmental costs: the exploitation of this resource requires the use of the hydraulic billing technique, which includes the injection of millions of litres of water mixed with chemical compounds that can potentially result in the contamination of aquifers (Yergin, 2011).

The combination of technological advances in hydraulic billing and horizontal drilling revolutionised the market and became known by Yergin (2011) as the Unconventional Gas Revolution, making resources known as shale gas,

[3] Williston Basin, Devonian-Mississippian, covering the central part of the USA to Canada.

coalbed methane (CBM) and tight gas; basin centered gas (BCG), which correspond, respectively, to the so-called gas in shale, methane in coal strata, and gas in low-permeability reservoirs. With these technological advances, natural gas has become competitively priced in the energy resources market, gaining advantages due to its lower pollutant content compared to other fossil fuels: coal and oil, and also gaining advantages over renewable energy sources due to its lower cost.

7.3.1. UNCONVENTIONAL NATURAL GAS

Since the so-called energy revolution in the United States, the exploration of natural gas from unconventional sources has become popular, making the world recognise this new exploration frontier and starting the internationalisation of this market (Yergin, 2011). In Brazil, studies and calculations of recoverable reserves carried out by the ANP in 2012 indicate promising figures for the country, with estimates of 200 trillion cubic feet of gas (Tcf) in just three Brazilian sedimentary basins: Parecis, Recôncavo and Parnaíba. Marco Aurélio Tavares, chairman of the board of directors of the consultancy Gás Energy, believes that this could be the opportunity to create a large national natural gas industry and a strong boost to Brazil's development (Petronotícias, 2012).

In general, there are three types of unconventional natural gas accumulations (Figure 9). The best known is the so-called shale gas, also known (although erroneously) as shale gas, popularised by the recent production *boom* in the USA. In addition to this, there is also CBM, which consists of methane generation from coal strata, and *tight* gas (BCG*)*, characteristic of low-permeability reservoirs. What defines these three types of natural gas accumulations as unconventional, apart from their low permeability, is an average recovery rate of 20 per cent, which is lower than the average for so-called conventional reservoirs, which ranges from 35 to 60 per cent. Therefore, it is necessary to implement different exploration techniques that favour the economic viability of unconventional deposits (Das Virgens, 2011).

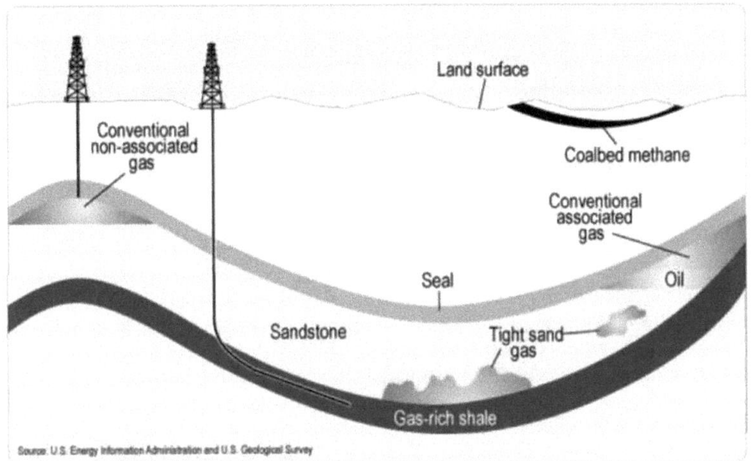

Figure 9 - Schematic representation of the main types of natural gas deposits. Source: U.S. Geological Survey (2012).

In the case of BCG and shale gas, they do not differ from conventional natural gas in terms of composition or origin, but only in the way they occur in the reservoir. In other words, it is predominantly methane, and can be of either thermogenic or biogenic origin (Law, 2002), concepts that are better discussed later in this chapter.

Natural gas is a mixture of light hydrocarbons in a gaseous state, known as the gaseous phase of oil. These gases can be isolated in a reservoir or associated with oil. The main component of natural gas is methane (CH4), and in many natural gas deposits the methane content represents approximately 92 to 95 per cent of the total composition. Various processes can result in the generation of methane, with the greatest production induced by bacterial degradation of organic matter. As the temperature of the system rises and the oil is consequently thermally altered, some of the constituent molecules are separated by heat, resulting in the formation of methane and other small hydrocarbon molecules. It can be said that the greater the depth of the system, the greater the rate of methane generation due to the higher temperatures. The repeated action of breaking down hydrocarbon molecules results in the formation of smaller and smaller carbon chains and, consequently, produces a greater number of molecules with associated carbons (from one to four carbons). These molecules correspond to natural gas, which is in a gaseous state at ambient pressures and temperatures. Therefore, the greater the depth of exploration drilling, the greater the tendency to find oil in the gaseous phase. Most commercial natural gas deposits follow this formation principle (Radovic, 1997).

In addition to methane, natural gas is also made up of small amounts of other hydrocarbon molecules (Table 2); these include ethane (C2H6), propane (C3H8), and butane (C4H10). Smaller quantities of pentane (C5H12) and hexane (C6H14) can also be found. Under normal pressure and temperature conditions, pentane and hexane appear in a liquid state, however, when exposed to high temperatures as a result of burial, they can constitute gases and make up part of natural gas. In addition to these hydrocarbons, natural gas deposits can contain inorganic components, including nitrogen, helium, carbon dioxide and hydrogenated sulphides (Radovic, 1997).

Table 2 - Representation of the average composition of natural gases in relation to hydrocarbon molecules. Analysis based on samples from reservoirs in Europe, Africa and North America.

Component	Vol.(%)	Component	Vol.(%)
CH4	81,5	N_2	5,2
C2H6	3,6	CO_2	4,6
C3H8	2,0	$H S_2$	1,4
C4H10	0,4	-	-
C5H12	0,2	He	0,04
>C5	1,0	-	-

Source: Neumann et al (1981) apud. Radovic (1997).

In terms of origin, natural gas can be categorised as biogenic and thermogenic (Figure 10). Biogenic gas is formed at low temperatures and buried depths (up to a maximum of 1,000 metres), under anaerobic conditions or associated with high rates of marine sedimentation. Around 20 per cent of the world's known natural gas is of biogenic origin. Thermogenic gas, on the other hand, is formed under higher temperature and burial conditions and is formed during middle and late catagenesis (the phase in which thermal methane replaces biogenic methane) as a result of the alteration of kerogen and other organic constituents through molecular breakdown and degradation. Thermogenic gas contains significant quantities of heavier hydrocarbons compared to biogenic gas. As the temperature rises, lighter hydrocarbons form as wet gas and condensate in the final stages of thermogenesis. Wet gas and condensates follow the progression of catagenesis followed by low-grade metamorphism (metagenesis) and thermogenic methane. As metamorphism progresses, oil is eliminated, resulting in an exclusive accumulation of natural gas (Tissot & Welte, 1978).

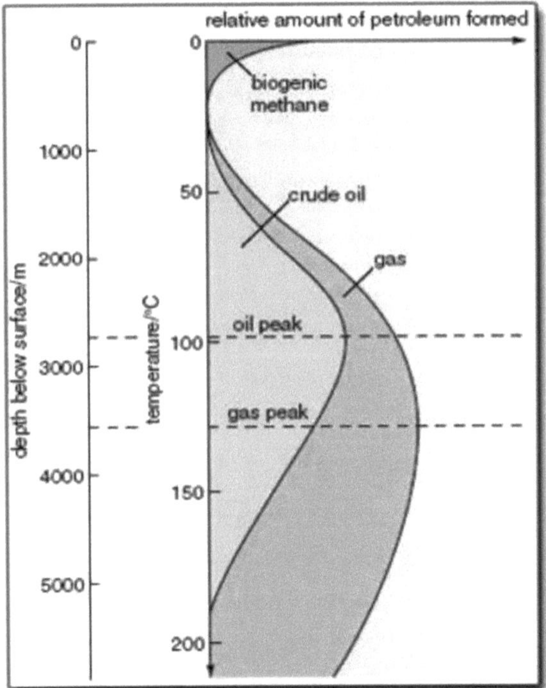

Figure 10 - Schematic representation of the relationship between temperature and depth of burial with the type and estimated quantity of oil generated.
Source: Modified from Tissot & Welte (1978).

7.3.1.1. *SHALE* GAS

The term *shale* gas refers to deposits of natural gas that are trapped in pelitic formations (shales). Shales are generally characterised by their low permeability, which makes it difficult to explore for oil from this lithology, previously considered a non-commercial source of hydrocarbons. However, the reduced permeability of shales, which makes it difficult to produce hydrocarbons directly from this formation, also acts as a factor in retaining natural gas inside this rock package (between the lamellae that accompany the foliation). Sheets that contain significant amounts of natural gas have the following properties in common: *composition enriched in organic matter (from 0.5 to 25%), and generally present in the field of thermogenic* gas (sufficiently high temperature and burial range, converting oil into natural gas - see figure 10, in the range below the gas *peak)* The gas generated from this process can be contained in natural fractures in the rock package, inside pores, and/or adsorbed to the organic-sedimentary material. The gas retained in pores and fractures is recovered during exploration activities, while the adsorbed gas is released as the reservoir's internal pressure decreases (Das Virgens, 2011).

Shale gas was first exploited in 1821 in the *Fredonia* region of New York, from surface fractures in layers of black shale. With the development of horizontal drilling techniques around 1930 and hydraulic fracturing in 1947, combined with the scarcity of oil and the decline in natural gas production during the 1970s, the exploitation of this type of gas became economic. Faced with declining natural gas production, the US federal government began to invest in the search for alternatives, creating the *Eastern Gas Shales Project.* The project began in 1976 and consisted of funding research in this sector, initially in the Michigan and Appalachian basins, as well as an industrial partnership, which received tax incentives from the government aimed at starting and expanding production. Since then, shale gas has been an important contributor to US energy generation. Shale gas has also become popularly known as shale gas[4] . However, this term is wrongly applied: shales are metamorphic rocks and are not considered hydrocarbon source

[4] It is believed that the term became popular in French literature, where it is referred to as *schiste gas, a* name that led to the mistranslation.

rocks (Yergin, 2011).

7.3.1.1.1. GEOLOGICAL ASPECTS

Shale is a sedimentary rock characterised by its fine granulation (< 1/256 mm) and fissility. It is predominantly made up of clay minerals, as well as smaller fractions of quartz and feldspars, and organic matter. These rocks have a high potential for generating hydrocarbons. In the case of shale gas, the generation and accumulation of natural gas occurs *in situ and* is concentrated in fractures and pores in the rock. These shales have a triple classification, being considered the source rock, reservoir rock and sealant of a petroleum system (U.S. Geological Survey, 2012). The migration process is very limited (primary migration), travelling only very short distances within the same rock package, in this case, the shale. These shales saturated in natural gas can also act as source rocks for other petroleum systems, with a higher disposition than these rocks, where the fluid will migrate to another lithology that plays the role of reservoir rock. According to several authors, the main factor favouring the concentration of this gas in shales is the overpressure stage of the system, reducing the rate of fluid migration and resulting in accumulations within the source rock itself, which also has sealing characteristics. In addition, the organic matter present in the system (abundant in shales) is capable of reabsorbing thermogenic gas molecules that have been generated from this shale, acting as traps for this gas within the system itself (ExxonMobil, 2012).

7.3.1.1.2. CURRENT SCENARIO

Shale gas has recently become an important source of natural gas in the United States and interest in this natural resource has expanded internationally. The combination of hydraulic fracturing and horizontal drilling exploration techniques has made it possible to exploit this resource, which was previously not considered economically viable due to technological limitations. In the 2000s, shale gas production represented just 1% of total annual natural gas production in the US, but by 2010 it had reached over 20% of the total, and according to the *Energy Information Administration* (EIA), production from this resource will reach 46% of total gas production by 2035. Currently, shale gas is the world's most exploited unconventional resource and accounts for more than 50 per cent of US unconventional natural gas production. According to studies and estimates, natural gas has played an important role in the energy sector and the electricity sector, corresponding to the second most consumed fuel in the United States today for primary energy and electricity generation (Figures 11 and 12) (EIA, 2013).

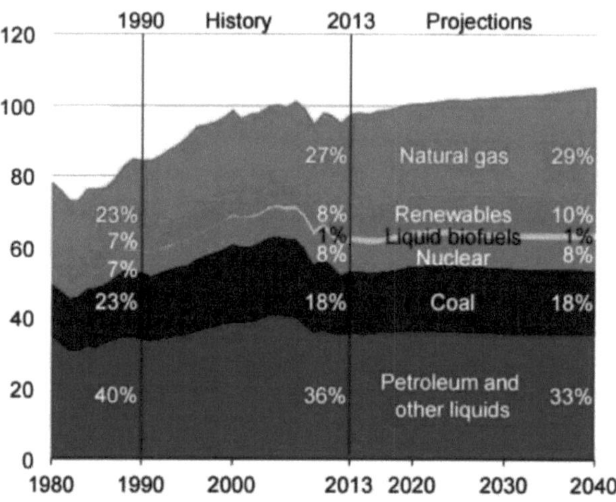

Figura 11 - Graphical projection of primary energy consumption by fuel in the United States. Source: U.S. Energy Information Administration (EIA), 2015.

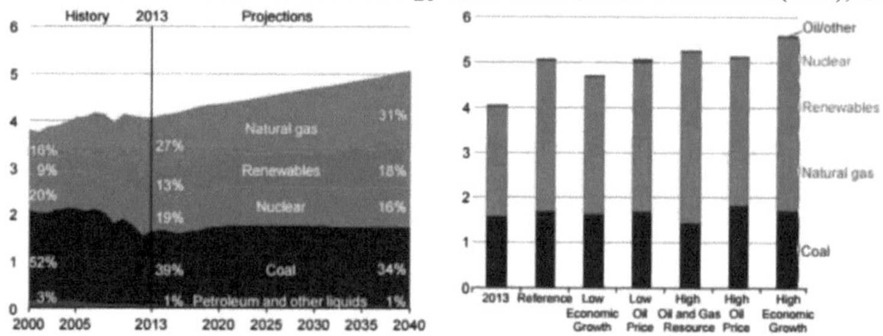

Figure 12 - Graphical projection of electricity generation by fuel in the United States. Source: U.S. *Energy Information Administration* (EIA), 2015.

In 2016, US President Barak Obama came out in favour of shale gas exploration and said that the development of this energy resource could help reduce the emission of greenhouse gases[5] . Recent studies indicate that, in general, the use of shale gas as an energy source results in the emission of fewer pollutants than conventional production (Forbes, 2012). However, other environmental factors associated with the techniques used to exploit this resource leave controversy over its environmental impacts, such as the contamination of aquifers and water bodies.

7.3.1.1.3. OCCURRENCE

Shale gas is found in portions of shales that are saturated with natural gas in significant concentrations. Its production began in the so-called *Barnett Shale,* located in Texas, and it has also been significantly exploited in the *Marcellus* Shale, located in the Appalachian Mountains region of the eastern United States (EIA, 2012). Through surface and subsurface geophysical data, combined with geological studies, other reserves of this type of gas have been found throughout the USA (Figure 13).

[5] Carbon dioxide emissions in the United States have been significantly reduced over the last five years (2010-2015). These reductions contrast with the estimates for CO emissions2 , made in 1998 as a result of the Kyoto Protocol. During the period of these discussions, the U.S. *Energy Information Administration* (EIA) had estimated that CO_2 emissions would increase annually at a rate of 1.3 per cent until 2020. It is believed that these estimates predicted an increase in pollutant emissions because they did not take into account the exploitation of cleaner, unconventional resources, such as shale gas, which was not yet significantly exploited at the time.

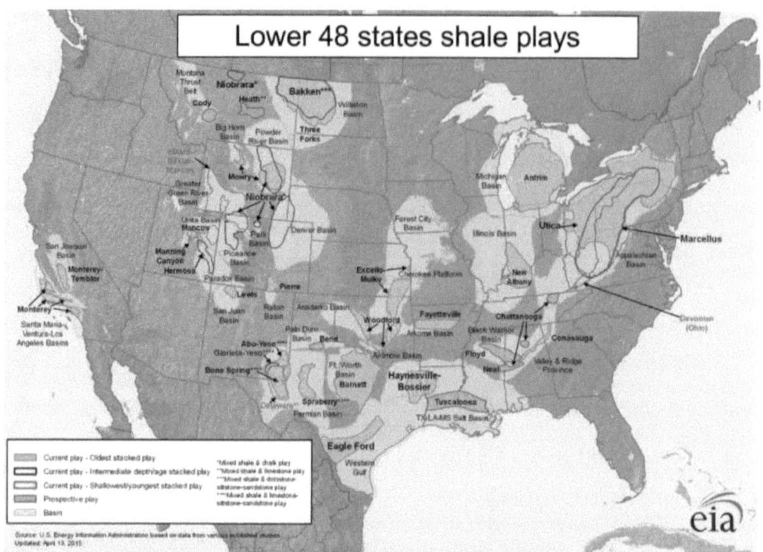

Figure 13 - Map of the North American sedimentary basins that produce shale gas. Source: U.S. Energy Information Administration (2015).

According to the EIA (2012) and as discussed in Chapter I, deposits of shale gas can be found on a global scale, with the largest concentrations in China (Table 3), with estimated values of 1,115 Tcf, followed by Argentina (802 Tcf), Algeria (707 Tcf), the United States (665 Tcf) and Canada (573 Tcf). Although several nations have the potential to produce shale gas, by 2013 only the United States, Canada and China were producing this resource on a commercial scale. Brazil is tenth on the list of countries with the greatest potential for this resource, estimated at 245 TCF. However, only 14 national reserves have had their potential proven, a fact that reflects the lack of national research in this sector.

Table 3 - Table showing the worldwide distribution of estimated and proven reserves of shale gas, organised according to the order of countries with the greatest potential for this resource.

Placement	Country	Estimated Recovery Values (Tcf)	Proven Natural Reserves
1°	China	1.115	124
2°	Argentina	802	12
3°	Algeria	707	159
4°	United States	665	318
5°	Canada	573	68

6°	Mexico	545	17
7°	South Africa	485	-
8°	Australia	437	43
9°	Russia	285	1.688
10°	Brazil	245	14

Source: Modified from U.S. Energy Information Administration, (2013).

To date, there is still no model for the development of gas in shale in Brazil. However, there is preliminary data on estimates made by the ANP, based on comparative studies of rocks and seismic data with basins in the United States. These studies generate calculations for recoverable volumes in the following Brazilian basins: Parnaíba Basin, Recôncavo Basin, Parecis Basin (Figure 14), with an estimated total of 245 trillion cubic feet of natural gas (Tcf) and the Solimões, São Francisco and Paraná basins, with unofficial estimates of 600 Tcf (ANP, 2013). In this context, it is believed that Brazil also has great potential for these unconventional accumulations of natural gas, and more studies are needed into this type of accumulation in order to discover new reserves.

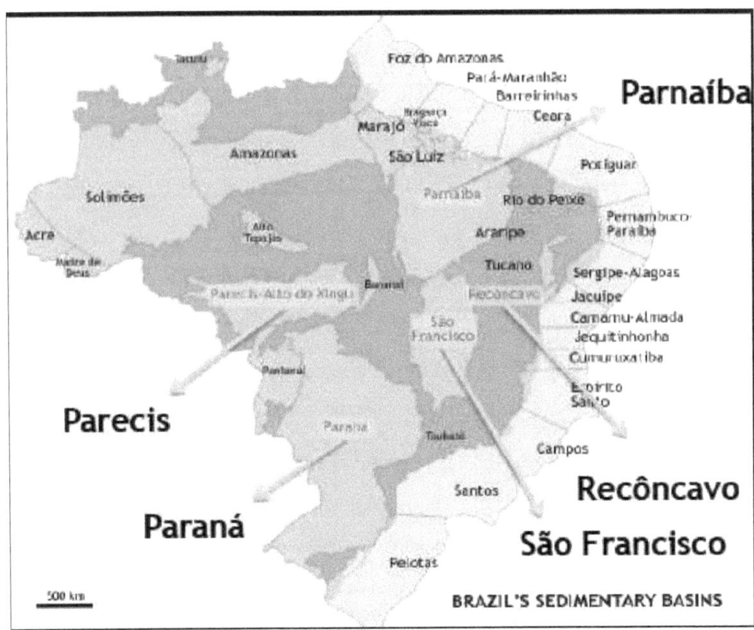

Figure 14 - Map of Brazil showing sedimentary basins with potential for shale

gas. Source: ANP (2013).

7.3.1.2.*COALBED METHANE (CBM)*

The so-called *coalbed methane* (CBM) corresponds to accumulations of natural gas from coal deposits. CBM is basically made up of methane that has been adsorbed to the solid matrix of the coal and is therefore trapped in this lithology. The occurrence of this type of gas is common in underground coal mines, where it can be released naturally, creating serious safety risks for operators, especially due to the threat of explosions. Methane released naturally during excavations is called *grisu* (French) or *firedamp* (English), with *grisu* being more common. The term CBM is reserved for the extraction of this gas from mineral coal strata. Unlike conventional sources of natural gas, CBM is characterised by containing small proportions of heavy hydrocarbons, such as propane and butane, as well as having no condensed phases, and is also classified as: (i) sweet *gas,* due to the absence of hydrogen sulphides in its composition, (ii) dry and non-associated[6] (Halliburton, 2008).

The gas that occurs in mineral coal is related to the original organic matter, i.e. Type III kerogen (vitrinite), which is basically made up of woody material (plant tissues, fibrous residues and humic fragments) with a low hydrogen content but a high oxygen content, with a greater potential for generating gas than oil. The gas generated is methane (CH4), but for standardisation purposes this gas is not called natural, but by its traditional name *grisu.* Although, based on chemical composition, there is no difference between coal gases and those associated with oil, the term natural gas only applies to oil gases. The formation of firedamp, or coal gas, takes place during the process of decomposition (putrefaction) of organic matter and the consequent thermal alteration it undergoes during the process of slow subsidence in the basin. The first members of the coal series have low gas adsorption capacity, but when the bituminous member is formed, the maximum

[6] The classification into *sweet* and *sour gas* is based on the concentration of H_2 S (relatively low for *sweet gas*, with values below 5.7 milligrams; and higher for *sour gas*). CBM is also characterised as non-associated dry gas, due to the absence of the liquid phase (oil) and condensate.

gas adsorption, both thermogenic (thermal alteration) and biogenic (decomposition of organic matter) is reached (Figure 15). Therefore, the volume of gas in the formation depends on the member of the series (Halliburton, 2008).

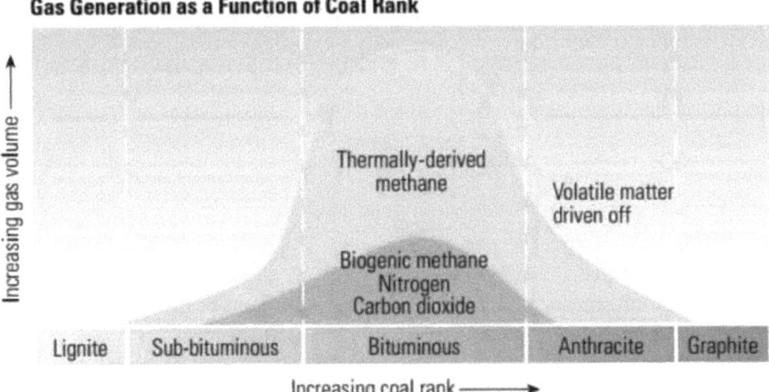

Figure 15 - Representation of natural gas generation in relation to coal classification and typology. Source: Halliburton (2008).

The term "*coalbed* methane" refers to methane gas that has been generated and trapped in strata of hard coal. Terms such as *coalbed gas* and *coal mine methane* can also be found in the literature to denote CBM. In Australia, a country with a high potential for generating this gas, it is called *coal seam gas,* known as CSG. Despite the existence of different terms, they all refer to the same type of natural gas deposit associated with coal strata. Attempts to make the extraction of coal seam gas viable gave rise to the English term *Coalbed Methane* (CBM). Over time it became common to apply the term directly to the energy resource (coal gas), but the term CBM refers to the extraction of the resource by pumping (gas) and not directly to the resource, which was already known and had its own name (coal gas or firedamp). Therefore, for all intents and purposes, a distinction must always be made between the resource and the form of extraction.

The same care must be taken when dealing with gas in coal as a gas reservoir. According to Halliburton (2008), these are gas reservoirs, albeit unconventional in different aspects, such as: the properties of porosity and permeability are partially active; the predominant property is adsorption. Based on this classification, the application of the concept of petroleum systems in a CBM context must also be adapted, as there are elements that make up the system: source rock and reservoir (coal strata) and migration, albeit restricted. In this context, the source rock adsorbs the product, becoming the reservoir itself. Thus, based on the process of generation and accumulation, and because the source and reservoir rocks are part of the same formation, the system can be classified as an independent petroleum system (IPS), or unconventional. Even so, the use of the term is questionable, as it could suggest that coal generates oil.

CBM exploration became popular on the American market in the 1970s, when the United States was experiencing a shortage of oil and a decline in natural gas production. As a result, research and investments were made in order to exploit this resource, which was abundant in the country due to the large use of coal as a source of electricity. Based on this scenario, the US Department of Energy set up research centres for unconventional sources of natural gas, including CBM, the production of which received tax incentives from the government, as well as shale gas and BCG (Halliburton, 2008).

7.3.1.2.1. GEOLOGICAL ASPECTS

CBM oil systems differ from other unconventional natural gas systems in that the main mechanism for gas accumulation is the adsorption process: instead of accumulating in the empty spaces between the grains (porosity), the CBM accumulates on the solid surface of the mineral coal, in micropores, due to the properties of its matrix[7] . In addition to this property, sedimentary rocks such as coal have low permeability, retaining the fluid in the lithology itself. This takes on the role of source rock, reservoir and sealant, as in the case of oil and gas systems

[7] The matrix present in sedimentary rocks such as coal is rich in organic matter. This has the capacity to reabsorb natural gas molecules that have been generated in the system. Therefore, natural gas molecules are associated with the coal matrix, accumulating on the surfaces of the system and characterising the property of adsorption, in a similar way to the situation described above for shale gas.

in shale. The generation of natural gas in coal occurs due to the continuous process of burying and compacting the package, resulting in higher pressure and temperature conditions, and consequent gas generation (Halliburton, 2008). The generation of CBM is associated with the process of thermogenesis and biogenesis. The methane produced in more primary stages of natural gas formation, such as lignites, comes from the process of biogenesis (biochemical decomposition), which consists of the breakdown of hydrocarbon molecules due to bacterial action. In addition to this process, thermogenic gas is also generated, associated with more advanced stages of thermal maturation (Figure 16).

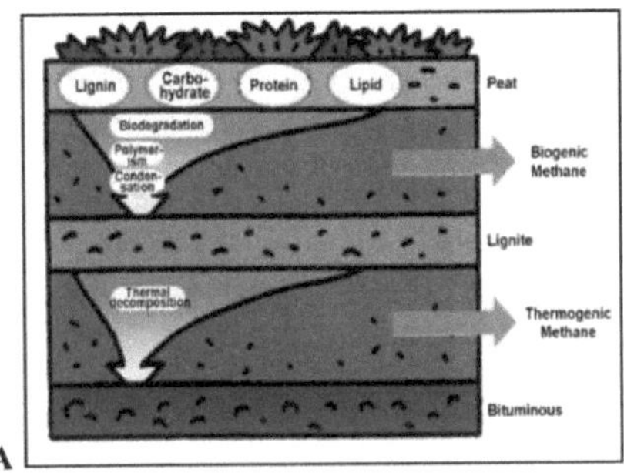

A

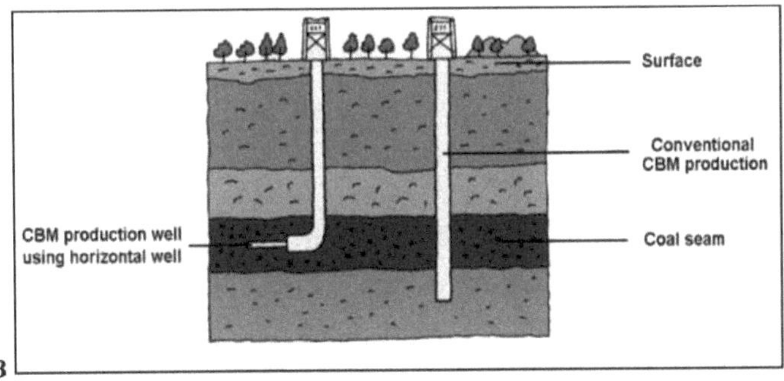

B

Figure 16 - Illustration of the arrangement of saturated coal seams in CBM. A: Representation of the stages of coal formation with the respective processes involved in the generation of biogenic and thermogenic gas (CBM). B: Representation of possible CBM exploration models, based on vertical drilling and horizontal drilling. Source: Sekitan no hon (2009).

7.3.1.2.2. CURRENT SCENARIO

Methane stands out among other fossil fuels due to its relatively low emission of pollutants and its low production cost. In a scenario of great demand for cleaner and cheaper energy resources, methane is a strong competitor in the fossil fuel market. In addition, methane deposits are abundant and widely distributed throughout the world, especially in the United States, where it was first exploited, being found in quantities of more than 800 Tcf in coal deposits. In this context, methane has shed its image as the villain of coal mines and has come to be seen as an environmentally friendly energy source (Halliburton, 2008).

According to Halliburton (2008), the exploitation of CBM brings numerous benefits to the oil industry by: providing a clean fuel with high energy potential (calorific value); expanding the world's natural gas reserves; increasing safety in coal mines, where CBM leaks can cause accidents; reducing the rate of pollutants emitted into the atmosphere from the combustion of fossil resources, and making mineral coal available for exploitation. However, the author disregards the environmental impacts related to coal mining and the contribution of methane to the greenhouse effect.

7.3.1.2.3. OCCURRENCE

The exploitation of natural gas from mineral coal strata has become an important tool in the search for energy resources to serve modern society. Financial market conditions and technological advances have made it possible to exploit this resource. Currently, CBM production represents an important part of the US economy: 10% of the country's annual natural gas production comes from coal strata (1.5 Tcf or 4.9 billion m^3), and approximately 9.5% of the country's total natural gas reserves correspond to this type of accumulation (Anderson et.al. 2003).

According to EIA (2012), as a result of the search for lower environmental impacts and lower costs, CBM production has received major investment from the American industry. The Rocky Mountain region in Colorado and the Appalachian

Mountains in the east of the United States are the main areas with CBM production; however, this resource can be exploited in any region with mineral coal reserves, such as Montana, Wyoming and other American states (Figure 17).

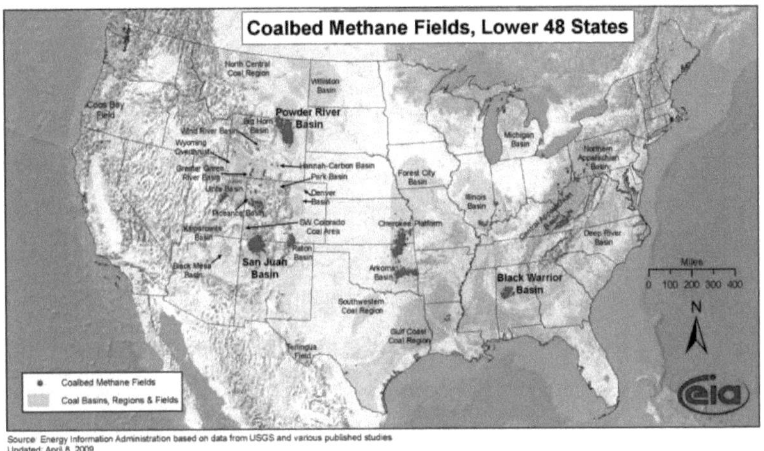

Figure 17 - Map of the United States showing the distribution and size of the main sedimentary basins that generate coal bed methane (CBM). Source: EIA (2015).

In addition to the United States, CBM is also significantly produced in the Queensland region of Australia and in the United Kingdom. Pilot projects have also been set up in countries such as China and India, as well as several other countries with large coal deposits where CBM is produced. The countries with the greatest potential are: United States, Spain, France, Poland, Australia, Canada, China, Great Britain, Germany, Zimbabwe, and Russia (EIA, 2012).

7.3.1.3. GAS IN LOW PERMEABILITY RESERVOIRS (BASIN CENTRED GAS - BCG & TIGHT GAS)

The concept of BCG began with Master in 1979, through publications on low-permeability sandstone reservoirs in the Alberta region of Canada (a definition currently attributed to tight gas). In his analysis of sandstones saturated in natural gas, which were referred to as gas in deep basins, Master (apud McCullagh & Hart, 2010) suggested that low permeability could act as a barrier preventing further fluid migration, the basic theory for BCG accumulations. These accumulations differ from conventional traps and are typically characterised by gas-saturated systems with anomalous pressures and low permeability. BCG accumulations vary from single (independent) oil systems, which are isolated and thin, to multiple systems with greater depth, differences which imply exploration strategies (McCullagh & Hart, 2010).

Due to controversial concepts and the scarcity of studies, BCG has been given various names over the years. The term tight gas is generally used in the literature as a synonymous definition of BCG. In many cases, the term BCG is not incorrect, however, it can generate ambiguous meaning and may refer to accumulations with conventional natural gas trapping (accumulations generated from the difference between density and thrust in the oil or gas column) rather than reservoirs with reduced permeability. In addition, the term tight gas tends to refer to compacted sandstone reservoirs, while BCG can refer to unconventional petroleum systems located in the portion of the reservoir where the gas is located. The term BCG refers to the positioning of the system and does not specify its lithology. The term deep basin gas can also be found in the literature, although its use is inappropriate: BCG accumulations are generally not found at great depths; it is assumed that the term refers only to accumulations at depths greater than 15,000ft (4,572m) (Law, 2002). Another term found is continuous gas accumulation, but this is a very generalised term and can also refer to shale gas. Due to the lack of consensus on the name, the most appropriate term, and the one chosen by this study as the official definition, is gas in low-permeability reservoirs, a term that encompasses both the definitions of tight gas and BCG.

7.3.1.3.1. GEOLOGICAL ASPECTS

BCG systems correspond to regional accumulations saturated in natural gas, with anomalous pressures (both low and high), usually without water-gas contact, and with reservoirs characterised by low permeability. In relation to the petroleum system, there are two types of BCG systems: a direct type and an indirect type, which can be differentiated based on the properties of their source rocks. The direct system has rocks with the potential to generate gas (gas-prone source rock, from type III kerogen), while the indirect system has the potential to generate oil (oil-prone source rock, from type I and II kerogen), a difference which influences various other properties such as thermal maturity and the pressure mechanism.

In addition to these two types of systems, there are hybrid systems in which both types of source rock (with oil and gas potential) contribute to the formation of the system (Law, 2000).

According to Law (2002), the development of a BCG system can be analysed through four reservoir pressure cycles (Figure 18). These cycles are determined from the dynamics of the geological processes involved and the system's response to these processes. The cycles represent evolutionary phases of the system in relation to the pressure involved, and will be described in detail below.

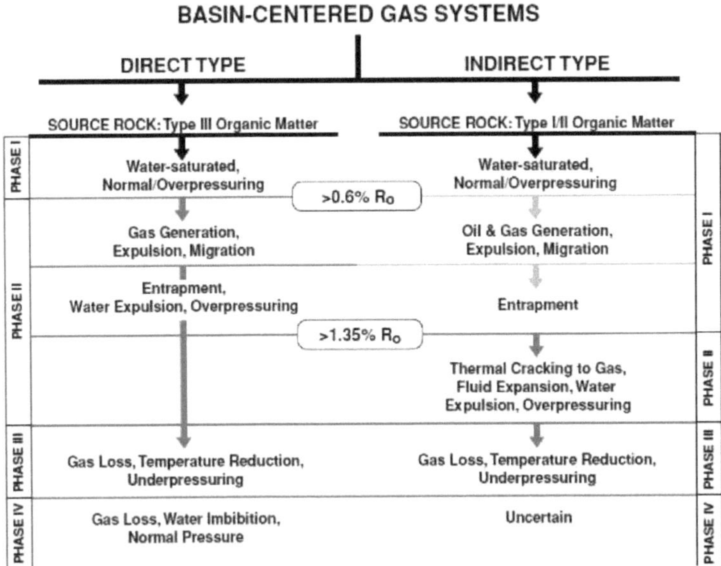

Figure 18 - Schematic diagram showing the evolution of direct and indirect Basin Centred Gas (BCG) or tight gas accumulation systems. Source: Law (2002).

- Phase I

During the beginning of the burial process and the thermal activity of direct and indirect systems, the reservoirs generally have normal pressure and the porosity of the rock is 100 per cent saturated with water. Compaction and grain arrangement are important elements in this process. The definition of phase I differs from direct to indirect systems: in direct systems, phase I ends with the generation of thermal gas, while in indirect systems, it ends with the start of thermal cracking of oil to gas. In addition, during phase I, indirect systems tend to have better quality reservoirs compared to direct systems because the accumulation of oil diapirs requires greater porosity and permeability than natural gas. In exceptional cases, reservoirs can present anomalous pressure during phase I: a situation generated by depositional events of rapid sedimentation, resulting in heterogeneous compaction and a consequent rise in pressure in the system. In these situations, the water in the system is equivalent to the fluid phase that exerts this pressure. However, as a result of the progression of the compaction and burial process associated with the heating of the system, the generation of hydrocarbons may replace the fluid phase and be responsible for the increase in pressure in the system. Examples of this transition in pressure mechanisms are the formations dating from the Miocene and Pliocene in the Bekes and Makó basins in Hungary (Law & Spencer, 1998).

- Phase II

Direct systems are made up of rocks with gas-generating potential in close proximity to low-permeability reservoirs. As the system is exposed to higher compaction rates and temperatures, the generation of natural gas begins, concomitant with the expulsion and migration of this gas into water-saturated sandstones and adjacent disposal. Because these reservoir rocks have low permeability, the rate at which gas is generated and accumulates in the pores is greater than the rate at which this fluid is lost. As a result, the pore capillary pressure is exceeded, releasing the water previously present in the system and resulting in high-pressure, gas-saturated reservoirs with little or no water in the system. Examples of this system include the Greater Green River, Wind River, Big Horn and Piceance basins in the Rocky Mountain region of the United States, and the Taranaki basin in New Zealand. In contrast to direct systems, indirect systems consist of rocks with the potential to generate oil, which is generated, expelled and accumulated in reservoirs from structural and/or stratigraphic traps in the form of diapirs, superimposed on the water present in the system. With subsequent burial and exposure to higher temperatures, the oil accumulations undergo thermal cracking and consequent gas generation, a process associated with a significant increase in pressure and fluid volume in the system. Under these conditions of high volume and pressure, there is also an increase in the capillary pressure of the saturated pores, expelling free water molecules and replacing the saturation with natural gas. In addition to the occurrence of this process, the system needs the presence of a sealing lithology for the initial accumulation of oil to be possible.

- Phase III

As BCG systems become exposed to overpressure conditions, the process related to the transition from phase II (overpressure phase) to phase III is identical for both direct and indirect systems. According to Law (2002), phase III occurs when an overpressured phase is exposed to reduced pressure conditions resulting from tectonic and/or erosional activity, as well as possible heat flow disturbances. Uplift events describe this situation, representing interruptions in the thermal history and the continuity of burial and compaction of the system. These events represent a reduction in temperature and pressure, as well as resulting in gas losses from the system. Studies show that there is greater gas loss in direct systems than in indirect systems, which is due to the integrity of the sealing rocks. Examples of direct systems with reduced phase III pressure are observed in the Cretaceous rocks in the San Juan, Raton and Denver basins in the United States; and as examples of phase III in indirect systems we have the Silurian rocks in the Appalachian basin

(USA), Ordovician rocks in Risha (Jordan), and Cambro-Ordovician reservoirs in the Ahnet basin (Algeria).

- Phase IV

According to Law (2002), phase IV is theoretical in nature and its concept is more applicable to direct systems, since sealing rocks are more effective in indirect systems than in direct systems. Phase IV corresponds to a continuous loss of gas from the system due to capillary pressure, associated with the slow and gradual re-entry of water into the gas-saturated system due to the reduced pressure. This

Thus, Law (op.cit.) suggests the hypothesis that these gas-saturated, low-pressure systems evolve into normal pressure systems, completing a cycle between pressurisation phases. Elements of BCG petroleum systems have specific characteristics that help differentiate direct and indirect systems. The nature of the source rock is the main factor that distinguishes the two systems. Source rocks in direct systems are generally represented by humic coal beds and carbonate shales, while indirect systems feature hydrogen-rich shales. The carbonate rocks in these systems have been described by Law (2002) as pervasive regionals that generally constitute isolated reservoirs of low thickness or vertical arrangement reaching great depths. For both systems, the reservoir rocks have low porosity (less than 13%) and low permeability (<0.1mD). They are composed of sandstones, siltstones and, rarely, carbonates. The depositional environment of these systems varies from marine to non-marine. Based on their morphology, BCG reservoirs can be divided into lenticular, such as deltaic channel sandstones, showing low pore volume and very low permeability; and sheet-like, such as interbedded river, deltaic front and eolian sandstones, showing higher pore volume and permeability compared to lenticular reservoirs. Distinguishing between these reservoir types is an important factor in the drilling phase. The sealing rocks of BCG systems are relevant for separating reservoirs between water-saturated and gas-saturated layers (figure 19). The nature of the boundaries between these fluids is determined by the integrity of the sealant, which can vary from lithological (low permeability rocks) to mechanical (from capillary pressure) (Law, 2002).

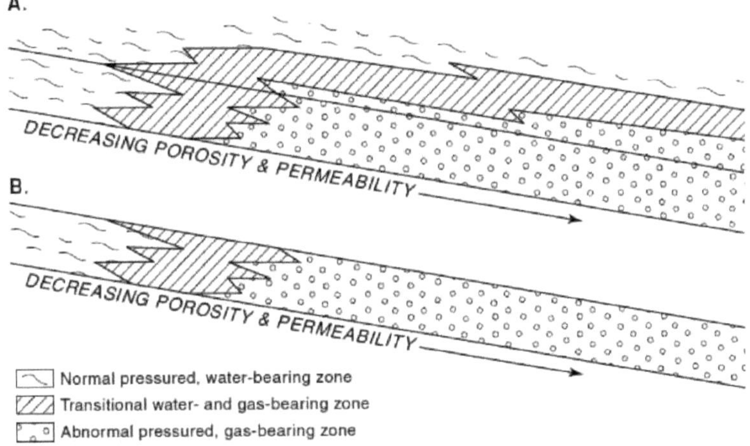

Figure 19 - Illustration showing transitional boundaries between normal pressure rock layers, saturated in water, and overpressure layers, saturated in natural gas, and their distinction into direct (A) and indirect (B) BCG systems. Source: Law (2002).

With regard to the processes at work in BCG-type petroleum systems, the generation of hydrocarbons and the change of physical state from oil to gas, due to the overpressure characteristic of these systems, occur under higher thermal maturation conditions. In general, hydrocarbon migration in BCG-type systems is carried out over short distances (immediate contact between source rock and reservoir), although exceptions can occur in cases of rutile deformation of the system, where the gas can travel longer distances from faults and fractures. The formation of traps is very important in indirect systems and is considered secondary in direct systems. In direct systems, the top of the system overcomes structural and stratigraphic limits, and the development of such traps is not necessary. In the case of indirect systems, conventional trapping is essential for hydrocarbons to accumulate, just as in conventional systems (Law, 2002).

According to data from Schlumberger (2013), BCG-type systems can be identified by analysing depositional environments and related diagenesis, as well as identifying independent reservoirs and the absence of contact between natural gas and confined water, in association with studies of the thermal history and compaction of the system so that the necessary conditions for the formation of this type of petroleum system occur. According to ExxonMobil (2013), the United States has been producing BCG for more than four decades and, due to its investment in research and technology, many deposits have already been identified and this resource currently accounts for around 40 per cent of national production of unconventional natural gas.

7.3.1.3.2. CURRENT SCENARIO

BCG systems are considered to be one of the most economically important unconventional systems in the world, contributing more than 40 per cent of total annual natural gas production in the United States (ExxonMobil, 2013). BCG exploration activity has been concentrated in North America, especially in the United States. In other parts of the world, BCG concepts are still little known and the exploitation of this resource is still poorly realised due to a lack of information, despite continuous progress. In order to develop and encourage this exploitation, it is necessary to study and break paradigms about petroleum systems, modifying traditional concepts and introducing new concepts about unconventional systems

7.3.1.3.3. OCCURRENCE

The occurrence of low-permeability reservoirs is widespread in the United States, especially in the Rocky Mountain and Appalachian Mountain regions, as well as in several other sedimentary basins that have the necessary conditions for the generation of BCG (EIA, 2012), as shown in Figure 20.

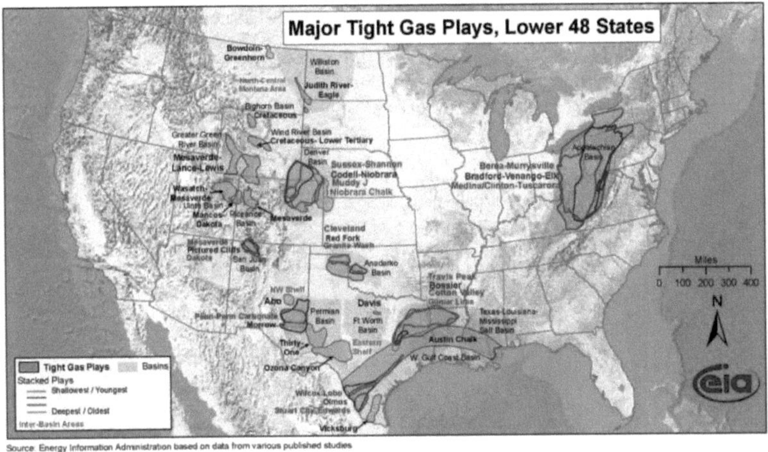

Figure 20 - Map of the United States showing the geographical distribution of known BCG deposits. Source: EIA (2015).

7.3.2. EXPLOITATION TECHNIQUES

The main objective of exploration activities is to define drilling areas. In the case of unconventional energy resources, exploration strategies generally take place in four stages: reconnaissance, confirmation, delimitation and identification of sweet spots (Law, 2002). These strategies are applied to the production of oil and natural gas and must take into account elements such as the source rock, reservoir, thermal maturation levels and the temporal relationship between these elements and the processes of generation, expulsion, migration and accumulation related to the petroleum system in question.

The Reconnaissance phase involves identifying sedimentary basins with the potential to generate unconventional natural gas. To do this, the nature of the source rock and the properties of the reservoirs (porosity and permeability) must be assessed. [8] Once a basin with oil potential has been identified, the Confirmation phase begins. As most unconventional gas reservoirs have anomalous pressures,

[8] In the case of BCG exploration, for example, the identification of coal seams saturated in thermogenic gas, associated with sandstone rocks with low permeability and porosity, is important for the Reconnaissance phase, acting as strong evidence of potential BCG concentration.

the confirmation phase of these systems evaluates the reservoir pressure values and the mechanisms responsible for the anomalous pressure, usually measured by indirect methods such as analysing the speed of sound waves emitted in the system. The system delimitation phase consists of the vertical mapping and spatial distribution of hydrocarbon accumulations. The delineation of accumulations is generally based on data acquired during the Confirmation phase, analysing anomalous pressure patterns. Finally, the sweet spot identification phase is carried out (areas of higher gas concentration in the system), since natural gas is generally not distributed evenly throughout the reservoir. The identification of these higher concentration portions is based on structural and stratigraphic aspects of the reservoir, as well as taking into account areas of anomalous pressure. Once the exploration areas have been defined, exploratory techniques are applied to enable the recovery of this resource (Law, 2002).

The low permeability characteristic of unconventional petroleum systems such as gas in shale, methane in coal strata and gas in low-permeability reservoirs is the main factor limiting the exploitation of natural gas from these systems. However, the development of exploration techniques such as horizontal drilling and hydraulic fracturing has made it possible to commercialise this unconventional natural gas.

7.3.2.1. HORIZONTAL DRILLING

The horizontal drilling technique, also known as directional drilling, corresponds to the process of drilling wells in which the initially vertical route is deviated, intersecting the reservoir at a practically horizontal slope. As most reservoirs are more extensive laterally than in thickness, the technique allows better access to the fluid concentrated in the reservoir, intensifying its recovery. In this way, horizontal drilling results in greater exposure of the hydrocarbon-saturated rock than the conventional vertical drilling technique, as it penetrates the reservoir perpendicular to the larger plane. The point at which the deviation from the vertical drilling route occurs is called the kickoff point, marking the moment when the borehole begins to intersect with angularity. The point at which drilling reaches the saturated rock is called the entry point, from which hydrocarbon exploration/production will take place (Figure 21) (Zou et.al. 2013).

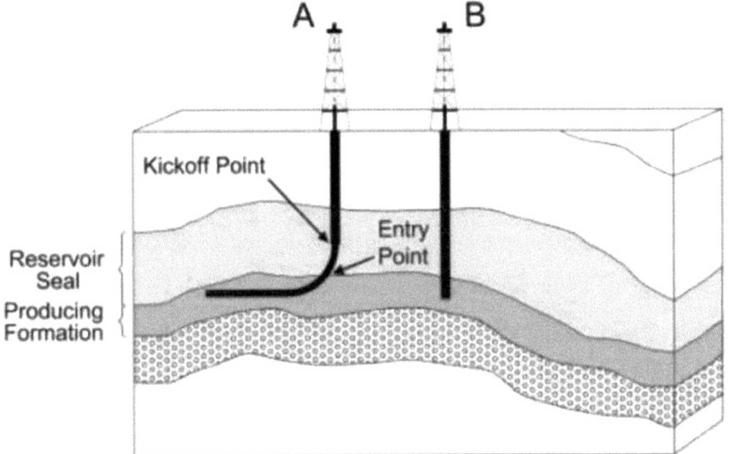

Figure 21 - Schematic representation of vertical and horizontal drilling techniques. A: Formation saturated in hydrocarbons being exposed to horizontal drilling, indicating the kickoff and entry points. B: Recovery through conventional (vertical) drilling. Source: EIA (2013).

The methodology of the horizontal drilling technique involves a rotating system made up of drill bits and drill pipes, standard for both vertical and horizontal techniques. From the kickoff point to the entry point, the curved section is drilled by means of hydraulically operated motors, which rotate the drill pipe and generate an inclination from the surface motor. Instrument packages attached to the drill pipe are responsible for transmitting sensor data to the surface operators. Generally, these sensors provide azimuth angles so that the system's inclination can be determined. More modern systems allow surface operators to identify the position of the drill bit (from x, y and z coordinates) at all times during drilling. The sensors also provide information about the drilling environment, displaying pressure and temperature values, rotation speed and the weight exerted by the rock package on the drilling rig, with the aim of avoiding or reducing wear and tear on the equipment (Schlumberger, 2012).

Horizontal drilling techniques offer several advantages. As well as increasing the contact surface between the reservoir and the exploratory well, the technique allows exploration in areas that were previously unreachable, such as reservoirs located under urban centres, bodies of water, among others. The technique also favours networked exploration, making it easier for other wells to be coupled to the drilling system and for the fluid to be explored in different directions, a process known as multilateral drilling (Figure 22). However, according to Schlumberger (2012), implementing this sophisticated exploration technique comes at a cost: a horizontal well can cost 300% more than conventional wells. Faced with this situation, the horizontal drilling technique is generally restricted to situations in which vertical drilling is not effective for recovering hydrocarbons[9], although it is very widespread in the United States (Figure 23).

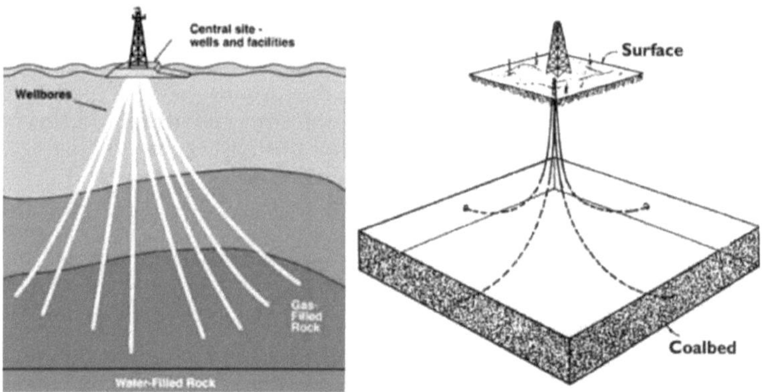

Figure 22 - Schematic representation of the network drilling technique, also known as multi-lateral drilling, where several drill pipes are coupled to the drilling centre in order to allow the fluid to be extracted in different directions during the same time interval. Source: EIA (2013).

[9] In oil reservoirs with good permeability in all directions and the absence of gas and water layers, the use of horizontal drilling is not essential, as conventional vertical wells can achieve the same level of reservoir recovery at lower costs. However, in the case of reservoirs with low permeability, or in cases where the presence of confined gas and water layers can interfere with recovery rates, horizontal drilling is economically viable, compensating for the cost of implementation with recovery values.

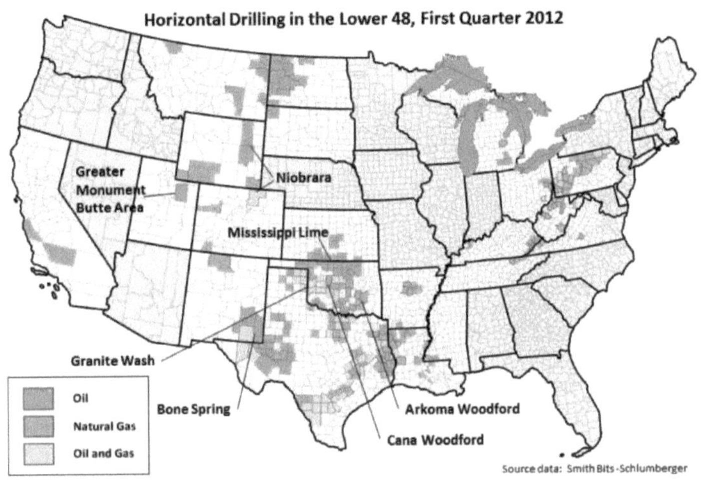

Figure 23 - Map of the United States showing areas where the horizontal drilling technique has been implemented. Source: Schlumberger (2012).

The first horizontal well in American history was drilled in the Texon region of Texas in 1929. However, the technique didn't become popular until the 1980s, when technology made this type of exploration economically viable. Even so, as a large part of exploration by horizontal drilling is carried out in unconventional reservoirs in the overpressure stage, the technique is not effective when used in isolation, requiring the implementation of another exploration technique that favours the flow of oil into the well, i.e. what is known as hydraulic fracturing.

7.3.2.2. HYDRAULIC FRACTURING

According to FracFocus (2010), hydraulic fracturing is a technique for recovering reserves once drilling has been completed. The technique involves injecting fluids under high-pressure conditions into reservoirs, creating small fractures in the formation in order to allow oil (or water, in the case of aquifer exploration) to pass through to the exploratory wells. The pressurised fluid used is basically made up of water, sand and small concentrations of chemical compounds. The release of the fluid into the system occurs in a similar way to small explosions, resulting in the fracturing of the system due to the high pressure. The sand fractions (and/or aluminium hydroxide grains) are responsible for keeping the micro-fractures open to maintain flow continuity. According to Schlumberger (2012), the size and quantity of fractures are decisive factors in boosting hydrocarbon production. These are classified as microfractures (generally less than 1 mm), through which the oil migrates to the well. The addition of chemical compounds (usually acids) is carried out to help initiate the process of microfracturing the rock due to the phenomenon of corrosion, as well as influencing the viscosity of the fluid in order to facilitate its penetration into the formation. The technique is quite common in shale gas, CBM and BCG explorations (Figure 24).

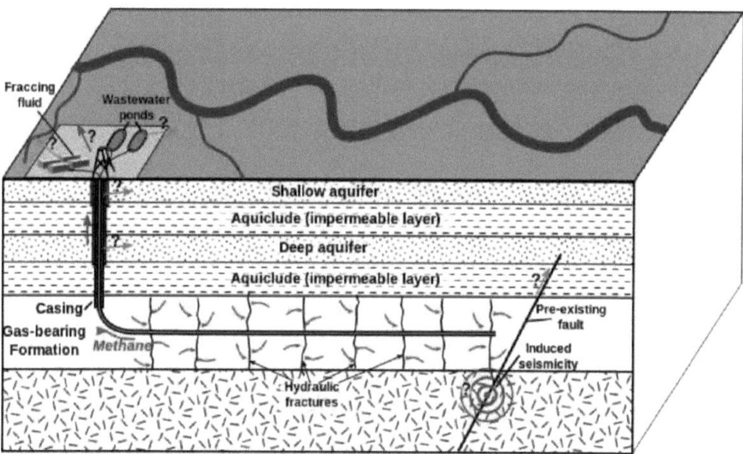

Figure 24 - Block diagram showing how the hydraulic fracturing technique works in conjunction with the horizontal drilling technique. Source: Fracfocus (2010).

Hydraulic fracturing is widely used to boost oil production in both conventional and unconventional reserves. However, the use of the technique is essential in unconventional deposits, where the exploitation of hydrocarbons only becomes economically viable due to the implementation of this technique (usually associated with the horizontal drilling technique) due to situations of overpressure in reservoirs. Investments and technological advances have improved the hydraulic fracturing technique, popularising it not only in the United States, where it is the main technique for recovering reserves (Figure 25), but also on a global scale.

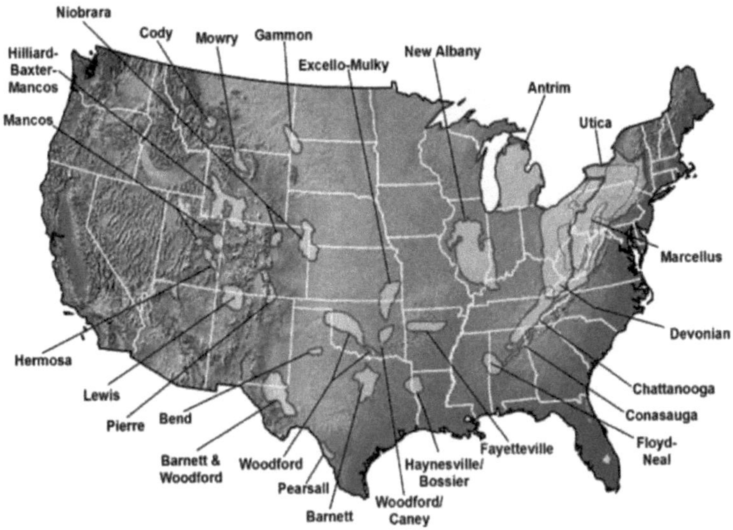

Figure 25 - Map of the United States showing the main areas where hydraulic fracturing is used. Source: EIA (2012).

The first hydraulic fracturing experiment was carried out in 1947, and the first commercially effective applications came shortly afterwards, in 1949. Researcher Floyd Farris was the pioneer in the use of hydraulic fracturing, but George P. Mitchell was responsible for the significant success of the technique's implementation in the 1990s, and is considered the "father of *fracking*". According to data from the EIA (2012), in 2010, 60% of the world's natural gas production used the hydraulic fracturing technique, replacing the use of explosives to recover hydrocarbons. Experts believe that in the next ten years, 80 per cent of oil wells in the United States will use hydraulic fracturing to continue operating (FracFocus, 2013).

The technique is still little known and not yet used in Brazil. However, according to the opinion of the ANP's 12ª Bidding Round, implementation of the technique was permitted for the exploration of unconventional natural gas, but was later vetoed until more detailed studies were carried out by means of a moratorium. Despite the need to use the hydraulic fracturing technique for the production of unconventional hydrocarbons, controversial opinions regarding the possible environmental impacts associated with it have generated controversy over its use.

7.3.3. ENVIRONMENTAL CONSEQUENCES

Although horizontal drilling and hydraulic fracturing techniques are part of the history of the oil industry in the United States, controversies regarding the environmental impacts caused by these techniques have threatened their popularisation on the world market. Public debates between the oil sector and environmentalists have been rife in the country and around the world. However, the veracity of these environmental disasters has yet to be scientifically proven.

The main potential environmental impacts involved are related to the contamination of aquifers and surface water bodies. However, according to Mandalios (2013), hydraulic fracturing can also result in the phenomenon of land subsidence, small seismic tremors and even collapse. The controversy surrounding the implementation of these techniques also encompasses the issue of intensive exploitation of these water resources: since the hydraulic fracturing process requires large volumes of water to be implemented, it could potentially interfere with the availability of water for urban supply. In addition, the addition of chemical compounds, also characteristic of this exploration technique, could act as a potential contaminant of groundwater and surface water (in the case of reverse flow in the system); a situation that is aggravated in the absence of adequate treatment of the residual fluid from the process (Figure 26) (Mandalios 2013).

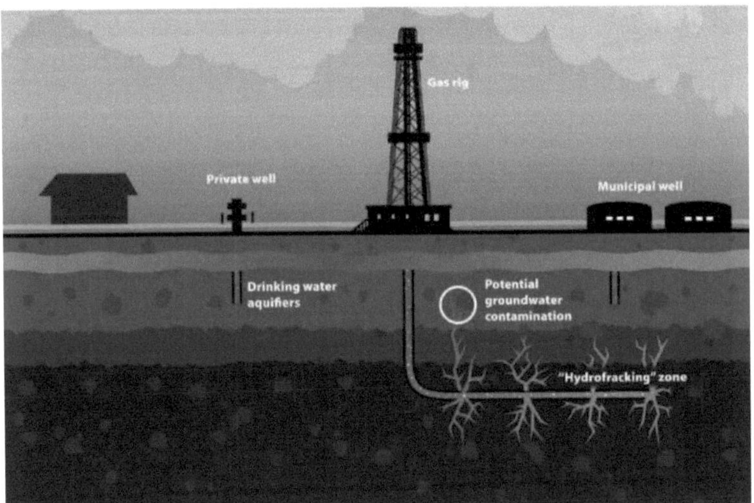

Figure 26 - Schematic representation of hydraulic fracturing and horizontal drilling techniques and their respective environmental impact related to groundwater contamination. Source: Mandalios (2013).

According to environmentalists, in addition to the intensive use of water, the most worrying factor is the chemical composition of the fluid used in the hydraulic fracturing technique. According to Werner et al (2014), fracturing fluid is made up of more than 600 chemical compounds, including concentrations of uranium, mercury, ethylene glycol, radium methanol, hydrochloric acid and formaldehyde, substances that are very toxic to the human body, have carcinogenic properties and are harmful to the environment. In addition, according to the author, only around 30 to 50 per cent of waste fluids receive treatment and are discharged into the ground, representing potential sources of contamination of groundwater (through infiltration) and surface water, as well as releasing toxic volatile compounds into the atmosphere. Faced with this situation, some countries have abolished the use of the hydraulic fracturing technique.

Hydraulic fracturing is the subject of controversy around the world. The method is permitted in Poland and the United Kingdom, but banned in France and Bulgaria. Other European countries have declared a moratorium on the extraction technique in order to carry out a more in-depth analysis of the potential environmental impacts, such as Ireland, the Czech Republic, Romania, Germany and Spain. The non-governmental organisation Greenpeace has officially come out against the method: it says it has serious concerns about the direct and indirect impacts on individual and public health. According to the organisation, many of these impacts are not just local, but can be felt at regional and even global levels (Deutsche Welle, 2013). However, it can be inferred that European resistance to the use of hydraulic fracturing may be associated with strategic and economic factors, such as the geological unfeasibility of these accumulations, the lack of investment and the required infrastructure, as well as the fact that North American leadership in the market potentially represents an economic barrier in the international oil industry; representing the usual scenario where environmental issues are subordinate to commercial policy.

In contrast to environmental organisations, the oil industry claims that the use of these techniques enables the exploitation of unconventional energy resources, such as shale gas, CBM and BCG, which represent cleaner sources of energy and reduce greenhouse gas emissions, such as carbon dioxide. In addition, to date, there has been no evidence of contamination of water resources through hydraulic fracturing and horizontal drilling; even if the potential for contamination is known, no significant concentrations of toxic components have been found (Werner et al., 2014). These are chemical compounds such as ammonium chloride, magnesium oxide, sodium chloride, boric acid, methanol, phosphoric acid, acetic acid, among others, which are also widely used for domestic and industrial purposes, such as in the cosmetics industry, in the production of detergents and other cleaning products, fertiliser and food production. In addition, this list of chemical products is known to be used for purposes such as: anticorrosives, friction reducers, fluid viscosity regulators, fluid thickness reducers, in preventing the precipitation of mineral salts and/or metals in the pipeline; in adjusting pH, in facilitating the flow of hydrocarbons, among other purposes essential to the oil production process; and which are progressively being replaced by organic compounds, in addition to obligatorily following standards and specifications for the implementation of this exploratory technique (Tassinari, 2015).

According to Forbes magazine (2012), in just five years the United States has achieved 70 per cent of the carbon dioxide reduction target set by the Kyoto Protocol in 1998. This is due to the advance in the exploitation of cleaner sources of energy (the boom in the exploitation of unconventional natural gas), which was not foreseen in 1998 (Figures 27 and 28).

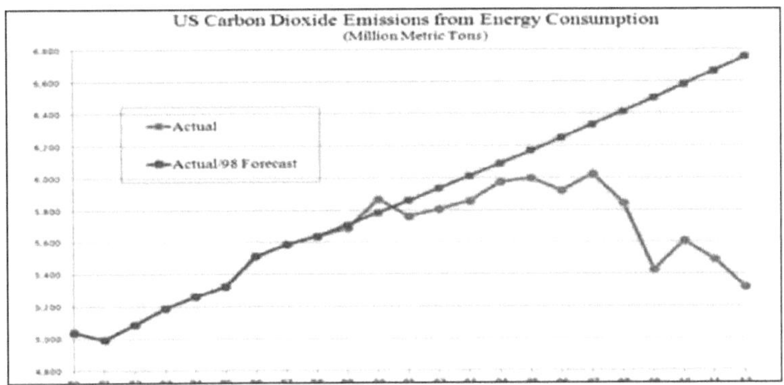

Figura 27 - Graph showing, in red, the carbon dioxide emission rate predicted in 1998 and the current emission rate, in green. Source: Forbes (2012).

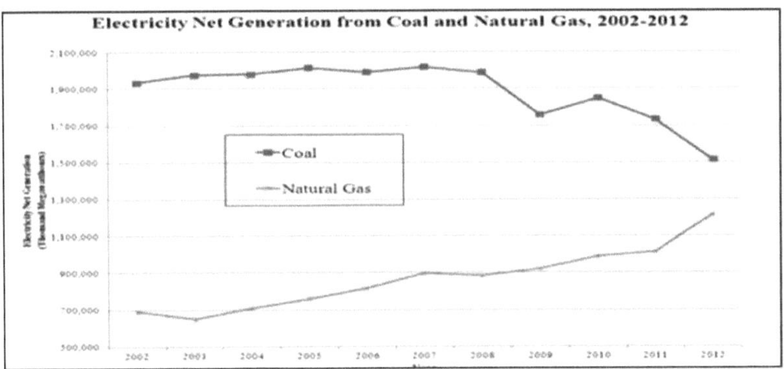

Figura 28 - Graph showing the recent decline in the use of coal as a source of electricity in the United States, and the increase in the use of natural gas as an energy source, a factor that implies a reduction in the emission of pollutants. Source: Forbes, 2012.

Faced with a society that is still very dependent on fossil fuels as energy generators, the environmental issue must be given considerable consideration. Despite the controversy surrounding the use of unconventional exploitation techniques due to their impact on health and the environment, the need to exploit these resources cannot be ignored. Faced with this situation, it is important not only to be environmentally aware and concerned about human health, but also to participate politically so that regulations are put in place regarding the implementation of these new technologies, the use of pollutants and the treatment of waste.

According to Radovic (1997), natural gas is an ideal fuel in terms of cost/benefit; it has a simple composition and high calorific potential (1000 BTU/ft3 or 24,000 BTU/lb), greater than any fossil fuel. Compared to other energy resources such as oil and coal, natural gas requires little processing before it can be used, as well as being easy to burn and emitting fewer pollutants than other non-renewable energy sources. The resource also requires less water consumption in its exploration and production, compared to other energy sources, as shown in Table 4:

Table 4 - Comparative use of water consumption (gallons) to generate energy (BTU) from different energy sources.

Energy Resource	**Water consumption (gallons/BTU)**	**Average water consumption (gallons/BTU)**
Shale gas	1-6	3
Nuclear (uranium enrichment)	8-14	11
Conventional oil	8-20	14
Synfuel (coal gasification)	11-26	18
Mineral coal	13-32	23
Shale oil	22-56	39
Bituminous sands	27-68	47
Ethanol (from corn)	2.510-29.100	15.805
Biodiesel (from soya)	14.000-75.000	44.500

Source: EIA (2014) apud Tassinari (2015).

In this context, sectors of the economy, industry and the environment are benefiting from the use of natural gas as an energy source. As a result, investments in studies and technologies have been made, not only in the United States, but around the world, with the aim of encouraging the exploitation of this relatively cleaner and more efficient resource.

CHAPTER 8

CHARACTERISATION OF FIELDS PRODUCING UNCONVENTIONAL NATURAL GAS

Organic geochemistry is a tool dating back to the early 1970s which uses chemical principles to study the origin, migration, accumulation, exploration and production of oil, as well as to help resolve environmental consequences such as contamination and pollution. Organic geochemistry can be divided into: molecular organic geochemistry, which is restricted to the bitumen fraction (soluble in organic solvents) through the analysis of biomarkers, aromatic compounds and polar compounds rich in nitrogen, sulphur and oxygen (NSO fraction); and non-molecular organic geochemistry, which characterises the kerogen fraction of sedimentary rocks. Geochemical characterisation techniques for hydrocarbon source rocks analyse the quantity, quality and level of maturity of the organic matter present. In order to obtain a more complete analysis, optical evaluation petrographic techniques are used, such as: vitrinite reflectance, liptinite fluorescence, alteration index and colouring of spores, acritarchs and microfossils; and geochemical techniques, such as: Rock-Eval pyrolysis parameters, total organic carbon (TOC) content, hydropyrolysis and biomarker characterisation. Source rocks can be classified according to the maximum amount of bitumen that can be extracted in dichloromethane, expressed in % (m/m) according to Table 5 (Azevedo da Silva, 2007).

Table 5 - Percentage ratio of bitumen and its classification in relation to the amount of organic matter present in hydrocarbon-bearing rocks.

Amount of Organic Matter	Organic Fraction or Bitumen (% by mass)
Poor	<0,05
Reasonable	0,05-0,1
Good	0,1-0,2
Very good	>0,2

Source: Modified from Azevedo da Silva (2007).

Once the amount of organic matter has been analysed, the initial parameter for classifying shales, the geochemical analysis is incorporated into the petrophysical and diagenetic analysis of the basin in general, continuing until the required level of detail is reached. For this study, based on bibliographic data from Andrews (2013) and the United States Geological Survey (USGS), the following criteria were determined to be essential in the classification and determination of *shale gas* producing fields:

Oil generation and preservation processes are mainly based on the level of maturation of the hydrocarbon and the type of organic matter contained in the source rocks (shales), parameters that are identified using the following geochemical characterisation techniques:

Total Organic Carbon (TOC%)

Among the geochemical characteristics essential to hydrocarbon generation, quantifying organic matter is generally the first analysis to be carried out to determine this potential in oil-bearing rocks. The process of measuring the amount of organic matter present in the rock is carried out through the total organic carbon (TOC) content, expressed as a percentage in relation to the dry stratum; a parameter which represents the conditions of formation and preservation of the organic content in the depositional environment (Azevedo da Silva, 2007). The amount of organic matter measured includes both the insoluble fraction (kerogen) and that soluble in organic solvents (bitumen). The experimental procedure for this analysis initially consists of treating the sample: it is pulverised and then diluted in hydrochloric acid in order to release the inorganic carbon content in the form of carbonate, generated from the interaction with the acid. The sample is then subjected to a combustion process, where the resulting CO2 content is measured, representing the organic carbon content of the sample. On average, TOC values suitable for hydrocarbon generation are around 2.0%, with minimum acceptable values of 1.0% for bituminous shales and 0.5% for carbonaceous shales (Table 6) (Azevedo da Silva, 2007).

Table 6 - Classification of the petroleum potential of source rocks according to TOC content (%).

Hydrocarbon Generation Potential	TOC (%)
Bass	0- 0,5
Medium	0,5-1,0
High	1,0-2,0
Very high	2,0-4,0
Excellent	>4,0

Taken from Azevedo da Silva (2007).

Rock-Eval Pyrolysis

The maturation rate of organic matter is another fundamental parameter for hydrocarbon generation, which is measured using the Rock-Eval pyrolysis technique. This technique simulates, under laboratory conditions, the natural processes of metagenesis and catagenesis that the rock has undergone, determining the stage of maturation of the organic matter present. The technique is initially implemented by overheating samples without the presence of oxygen so that combustion does not occur. The samples are subjected to temperature ranges between 25 and 600°C. During this heating process, free hydrocarbon fractions present in the samples are released, as well as carbon dioxide (CO2), both generated as a result of the thermal cracking of kerogen. The rate of release of these gases is measured using ionisation detectors, which record this release using a pyrogram in which three different peaks are observed: S1, S2 and S3, which will be analysed in geochemical studies (Figure 29).

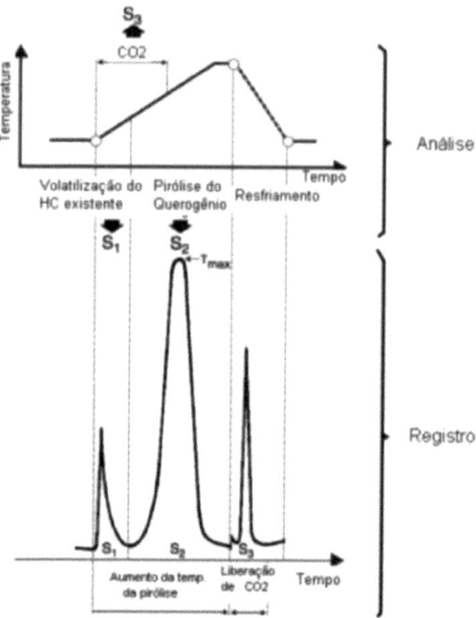

Figure 29 - Representation of the analysis cycle and example of a pyrogram obtained from Rock-Eval analysis. Source: Azevedo da Silva (2007).

The S1 peak represents the quantity (in mgHC/g rock) of free hydrocarbons (bitumen), the soluble fraction of organic matter in organic solvents. Peak S2 represents, in the same unit, the amount of hydrocarbons released by the thermal cracking of kerogen, thus being called the generating potential; in other words, the amount of oil that the analysed rock would be capable of generating if subjected to the appropriate conditions of burial, pressure and temperature. Peak S3 corresponds to the amount of carbon dioxide (CO2) released during the cracking process (in mg CO2/g Rock). In addition to measuring the peaks, related interpretations are also required: with the progressive conversion of kerogen into oil, there is also a progressive reduction in the generating potential (S2) and an associated increase in the amount of free hydrocarbons (S1), which is why the S1/(S1+S2) ratio, known as the Production Index (PI), is used as an indicator of the progress of the generation process and the level of maturation of the organic matter present in the rock (Azevedo da Silva, 2007).

Based on the results of Rock-Eval pyrolysis, a semi-quantitative assessment of a rock's generation potential can be made, with S1 representing the rock's initial generation potential (original generation fraction) and S2 representing the residual potential. In this way, the ratio S1+S2, expressed in mg of HC/g rock, corresponds to the total hydrocarbon generation potential of a rock, based on the type and quantity of organic matter present. Based on this analysis, the following classification of source rocks was determined (Azevedo da Silva, 2007): rocks with a generating potential (S1+S2) of less than 2 mg of HC/g of rock (< 2000 ppm) are not considered oil-generating rocks, but have some potential for natural gas; rocks with a potential of between 2 and 6 mg of HC/g of rock (between 2000 and 6000 ppm) are considered rocks with moderate potential for generating oil; and rocks with a potential of more than 6 mg of HC/g of rock (> 6000 ppm) are considered oil-generating rocks, with good generating potential.

In addition to evaluating the S1, S2 and S3 peaks, the temperature at which the maximum hydrocarbon generation occurs from the cracking of kerogen is also evaluated, reflecting the maximum thermal evolution that the rock has undergone, a property called Tmax. In general, samples with Tmax values in the range of < 435 to 440°C are considered immature rocks and have a Production Index (PI) < 0.1; while mature rocks, which have reached peak generation values, have Tmax values between 445 and 450°C and PI in the order of 0.4 (Azevedo da Silva, 2007).

Other parameters can also be determined from Rock-Eval pyrolysis, such as the Hydrogen Index (HI) and Oxygen Index (OI), which are analysed in order to determine the type of organic matter present and its degree of preservation: in general, organic matter of terrestrial origin is poorer in hydrogen and richer in oxygen, while algae organic matter is generally rich in hydrogen. Type III kerogens, essentially terrestrial, are characterised by low HI values (generally < 300 mgHC/gCOT) and high IO values (generally over 100 mgCO2/gCOT). On the other hand, type II kerogens, of marine origin, have HI values in the order of 400 to 700 mgHC/gCOT, and type I, of lake origin, have HI values generally higher than 600-700 mgHC/gCOT; both types II and I have low OI values (lower than 100 mgCO2/gCOT). These values correspond to immature source rocks, since throughout the maturation process and Rock-Eval analysis, IH and IO values tend to zero (Azevedo da Silva, 2007). In this context, the three basic types of kerogen (types I, II and III) can be identified by integrating this information: IH is obtained from the S2/COT ratio and expressed in mg HC/g COT, and IO from the S3/COT ratio and expressed in mg CO2/g COT (Azevedo da Silva, 2007).

8.1. Parameters adopted

The research method for this master's dissertation includes a survey of comparative geological parameters established by the U.S. Geological Survey (USGS) and other sources in the literature for determining shale gas producing fields (shale plays) (USGS, 2013; Andrews, 2013; Azevedo da Sila, 2007); with the aim of establishing which parameters are most appropriate and which will be adopted in this work to analyse the geological viability of the Paraná Basin for the occurrence of this resource. In this context, data was collected on the geology, organic geochemistry and petrophysics of shales, which were summarised in a table and are the basis for the comparative study between the main shale gas provinces.

8.1.1. Geological Parameters

Mineralogy and Clay Content (%): The clay content should be less than 35% to facilitate hydraulic fracturing. According to Andrews (2013) the silica content should be higher than 30% and carbonate fractions should also be present.

Layer thickness: Moderately thick shale layers are considered ideal (over 50 ft or 15m). It was thought that greater thicknesses were favourable for shale gas accumulations, however, studies carried out by the USGS show that very thick layers tend to result in BCG (basin centered gas or tight gas) type accumulations rather than shale gas.

Depth: The depth at which the shale layer is found must be greater than 5000ft (or 1500m). Lesser depths generally tend to result in reduced flow in natural gas production due to the pressure gradient.

Basin tectonics and diagenesis: Preferably large-scale (extensional) and stable (intracratonic) sedimentary basins, without high tectonic complexity throughout their evolutionary history (low degree of fracturing and faulting). Producing wells should be drilled far from faults and fractures.

8.1.2. Geochemical Parameters

Organic carbon content (% TOC): According to Andrews (2013), suitable total organic carbon (TOC) values for unconventional natural gas in shales should be higher than 2% (Charpentier & Cook, 2011), 4% (Lewis et.al.2004), or even 1% (Jarvie, 2012). According to USGS (2013), the average among the main shale gas producing fields in the United States is between 0.93-5.34% TOC.

Type of Kerogen: The kerogen must be type I or type II. Type II being more suitable according to Andrews (2013), USGS (2013).

Hydrogen Index (HI): should preferably be higher than 250 mg/g. Andrews (2013) emphasises the importance of the original HI data and not recent values, which mischaracterise the analysis.

Thermal maturity/geothermal gradient: The shale must be thermally mature for natural gas (thermogenic gas window), with ideal vitrinite reflectance values (Ro) of between 1.1 and 3.5 per cent or preferably between 1.75 and 3 per cent.

8.1.3. Petrophysical Parameters

Gamma-ray values: High gamma radiation is indicative of high organic matter content. Thus, high gamma ray values are ideal for the occurrence of shale gas: typically between 150 and 200 API.

Porosity (%): Typically between 4 and 7%. Necessarily less than 15%.

Gas saturation: Natural gas must be present both free (in the matrix and fractures) and adsorbed to organic matter. The ideal saturation is between 60 and 200 bcf per geological session.

CHAPTER 9

DESCRIPTIVE-COMPARATIVE GEOLOGICAL MODEL

Based on studies of oil fields proven to generate and produce unconventional natural gas and an analysis of the factors that condition the occurrence of this resource, the geological, geochemical and petrophysical parameters considered essential to this condition were surveyed and combined. This compilation is shown in the following descriptive-comparative table for determining the potential for unconventional natural gas in shales (Table 7).

Table 7. Descriptive model of the main parameters analysed in the geological feasibility assessment for the occurrence of unconventional natural gas accumulations in shales.

Comparative Parameters
Geological Age
Lithology Mineralogical composition (% by volume) Mineralogy (% volume) Area (km)^2
Thickness (m) Depth (m)
Predominant Kerogen Type (%)
TOC (%)
Tmax (°C)
Thermal Maturity
Sedimentary Basin Classification
Depositional palaeoenvironment Reservoir characteristics (%)

Source: Modified from Azevedo da Silva (2007), Andrews (2013) and USGS (2013).

Once the parameters conditioning the occurrence of unconventional natural gas had been identified, they were analysed in the main shale gas producing fields: the US Marcellus, Barnett and Eagle Ford fields. These fields were chosen for study due to their importance in the production of this resource, as they were the largest producers of unconventional natural gas between 2008 and 2014 (Norton Rose Fulbright, 2013) and pioneers in the production of this resource on a commercial scale, as well as being adopted as a reference for geological, geophysical and reservoir engineering standards for determining shale gas fields by the United States Geological Survey (USGS). In this context, these main producing fields were analysed in order to develop comparative studies between them and potential Brazilian fields by means of analogies.

8.2. Main Fields Producing Unconventional Natural Gas

- .1.1 MARCELLUS SHALE

The Marcellus Field belongs to the formation of the same name and is located in the north-eastern part of the United States, in the Appalachian Mountains (Figure 30). The Marcellus field has a recoverable potential of between 30 and 300 Tcf of natural gas, combined with the advantage of being close to the country's major consumer centres. The lithology consists of organic-rich black shale dating from the Devonian (approximately 390 Ma), and its deposition is associated with the Acadian orogeny (Lee et.al., 2011).

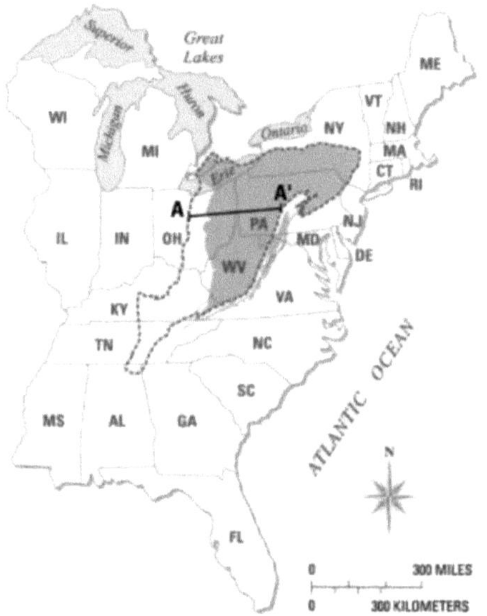

Figure 30 - Representative map of the geographical layout of the Marcellus geological formation, Appalachian Basin. Source: Lee (et al., 2011).

The Marcellus Shale was originally recognised as a hydrocarbon source

rock within the conventional petroleum systems of the Upper Devonian and Lower Carboniferous (Mississippian), which have been exploited since 1821. Hydrocarbon exploration in the Appalachian Basin, where it is located, was intensified between 2004-2005 due to the boom in unconventional natural gas production in the United States (shale boom). Based on the exploratory aspects used in the Barnett field in Texas, Marcellus exploration was initially successful, with the recovery of considerable volumes of hydrocarbons, mainly natural gas in the eastern and north-eastern portions of the basin, where the main shale gas production field in the region is located (West Virginia, Pennsylvania and New Yrok states) (Bruner &
Smosna, 2011) (Durham, 2008). This shale play is bounded by the Allegheny Formation to the east, the Adirondack Formation to the northeast and the Waverly Formation to the west. In the central portion of the producing field, the Marcellus formation is up to 50 feet (approximately 15 metres) thick and extends for approximately 50,000 mi^2 (129 km^2), the portion with the greatest potential for hydrocarbon production and generation. Due to the location of these portions, exploration, production and regulatory activities are mainly concentrated in the states of West Virginia, Pennsylvania and New York.

The Appalachian Basin is characterised as an asymmetrical foreland sedimentary basin formed over 200 million years. Throughout its formation period, the basin was subject to three different orogenies: Taconic Orogeny, dating from the Upper Ordovician and representing the collisional process between the Laurentia and Laurussia tectonic plates and a system of oceanic island arcs; Acadian Orogeny, from the Devonian to the lower Mississippian and resulting from the collision between Laurentia and a smaller-scale continental plate (Avalonia, Baltica or Armorica); and Allegheny Orogeny, from the Pennsylvanian to the Permian, generated from the convergence between Laurentia and Gondwana and the consequent formation of Pangea. The palaeogeographical reconstruction of the basin indicates that it was submerged in the portion that is now equivalent to the state of New York to Georgia, which promoted the accumulation of black shales throughout the basin, with the Marcellus shale being the oldest and most deeply located in the basin (although it is locally overlain by the Needmore shale). Since its deposition and burial, the Marcellus has been subjected to intense tectonic deformation activity in the basin, mainly associated with the Allegheny Orogeny. The Appalachian Basin is currently arranged in a northeast-southwest direction and its structural axis is at a depth of around 6,000 feet below sea level. Strata from the Devonian period, such as the Marcellus Shale, have an eastward structural dip, averaging 55 feet per mile from the Cincinnati Arch in Ohio, and are at a depth of about 10,000 feet (about 3,000 metres) in central Pennsylvania (Bruner & Smosna, 2011).

The Marcellus shale, together with the Mahantango sedimentary formation, make up the Hamilton group, dating from the middle Devonian (Eifelian and Givetian periods). The shale varies in colour between grey, dark brown and black, with a carbonaceous composition and intercalation with thin layers of limestone and carbonate concretions, as well as a high degree of radioactivity due to its rich content of organic matter. The limestone layers are fossiliferous and the shale is mineralogically rich in pyrite, especially in the lower portions of the formation. Its thickness varies between 140 and 660 feet (660 feet in the northeastern portion of Pennsylvania, which drastically reduces to 200 feet in the western and southwestern directions of the state, and to 140 feet or less in West Virginia and Ohio; respectively 200, 60 and 42 metres thick) (Boyce, 2009 apud Bruner & Smosna, 2011). The geographical limits of the Marcellus Formation are erosional in all directions (Figure 31) (Bruner & Smosna, 2011).

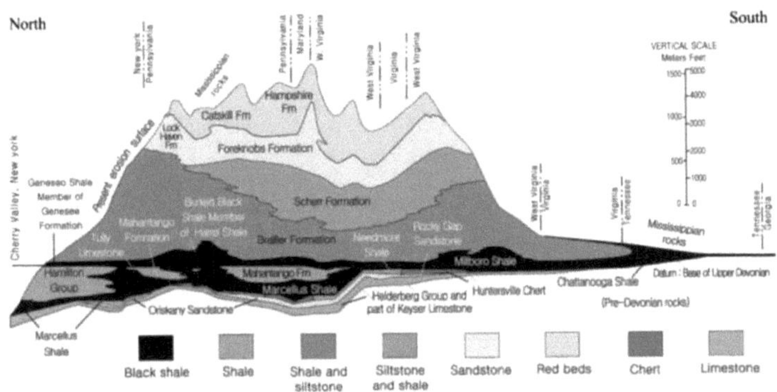

Figure 31 - Representative stratigraphic section of the Marcellus geological formation, Appalachian Basin. Source: Lee (et al., 2011).

Based on the analysis of outcrops carried out by Bruner & Smosna (2011), samples of Marcellus shale collected in western New York indicate the subdivision of the formation into three different stratigraphic members: Union Springs, Cherry Valley and Oatka Creek. The lower member, Union Springs, corresponds to grey to black shales that are characterised by their slight foliation, composition rich in pyrite, carbonate and organic matter. Strata of siltstone are also present among the layers, mainly in the lower portion of the package. Based on an analysis of isopach maps, the thickness of the member was determined to be around 60 feet (18 metres), which can reach 150 feet (45 metres) in northeastern Pennsylvania and 20 feet (6 metres) in Ohio. The intermediate member, Cherry Valley, corresponds to a fine-grained carbonate with intercalations of calcitic shale. Barite nodules are present locally. The formation is also called the Purcell, initially described in West Virginia, but some authors consider them to be different units. The thickness of the member varies from 50 to 100 feet (15 to 30 metres). The upper member is called Oatka Creek and is subdivided into two subunits: a basal subunit of black shales (similar to the Union Springs member, but with different gamma ray values); and an upper subunit of grey shales. The thickness of the Oatka Creek member, based on isopach analysis, varies from 270 feet (which can be as high as 570 feet or 82 to 173 metres) in northeastern Pennsylvania to 120 feet (36 metres) in Ohio, as well as 30 feet (9 metres) in central New York. According to Bruner &

Smosna (2011), this division of the Marcellus Formation represents the complexity of intercalations between organic-rich shales, which vary in composition between silty, carbonate and sandy fractions, as well as variations in the content of organic matter present.

Factors such as the depth of the formation and its low permeability classified the Marcellus as an unconventional petroleum system, whose production was not economically viable. However, the increase in the price of natural gas since 2008, coupled with significant technological advances, has led to the production of hydrocarbons from this geological formation through the implementation of the exploratory technique of hydraulic billing (Harper, 2008).

Marcellus shale has a high organic matter content, with total organic carbon (TOC) values of up to 20 per cent. However, the formation has TOC values predominantly in the order of 2 to 6%; type II kerogen is the predominant one (Bruner & Smosna, 2011). On average, the formation is made up of black shales enriched in organic matter, with the presence of calcite and pyrite and sparse fossiliferous content; interspersed with layers of siltstones and claystones. Analyses carried out on the black shales indicate values in the order of 35% clay minerals, 10 to

20% quartz/silica, values of less than 10% for feldspars, 5 to 30% muscovite and chlorite, 25% calcite, as well as pyrite and organic matter in its composition. According to Roen (1984 apud Bruner & Smosna, 2011), the formation has higher silica contents (between 40 and 60 per cent) in the eastern part of the basin

The Marcellus Formation covers a geographical area of approximately 140,000 km^2 (75,000 mi^2) and is located at an average depth of 2 km along the Appalachian sedimentary basin in the northeast of the United States, belonging to the states of West Virginia, Virginia, Ontario, Pennsylvania, Ohio, Maryland and New York. The formation is, on average, 30 metres thick, reaching values of around 75 metres in the northeastern part of the state of Pennsylvania. According to Williams (2012), the formation has an estimated potential of between 295 and 2,700 TCF of natural gas, with recoverable reserves estimated at around 10% of this figure; which corresponds to between 1 and 10 years of guaranteed natural gas supply in the USA, based on current consumption (21 TCF per year). According to Lee et al. (2011), both the thickness and depth of the shale increase towards the east of the basin, and the portions of the shale with the main concentration of natural gas are located in north-central Pennsylvania and south-central New York, with depths of around 60 metres, while shallower portions are found in the state of Ohio, with depths of less than 0.9 km. Although greater depths are generally associated with a higher concentration of natural gas, this condition also implies an increase in production costs. According to the authors, the formation's considerable potential is still associated with a number of technological and economic challenges, which include adequate prospecting and targeting the portions with the highest hydrocarbon saturation, and the effective recovery of this in the face of greatly reduced porosity and permeability properties.

According to Harper's classification (2008), the Marcellus fault has low porosity, in the order of 0.5 to 5.0%, although values in the order of 9.0% have been recorded in West Virginia. Due to the low connectivity between pores in the matrix of this shale, the recovery of hydrocarbons from this formation is closely associated with its fracture porosity, which varies between 2.0 and 7.0%.

The Marcellus Formation has two sets of fractures (J1 and J2), J1 has a preferential east-northeast orientation, parallel to the zone of maximum horizontal stress, while J2 has a northwest direction. According to Engelder et.al.(2008a), based on an analysis of the orientation of these fracture families, the direction of the directional drilling was determined in the direction of the J1 set of fractures, because they are more developed along the shale and have less spacing between them, and are therefore more permeable than J2, providing greater recovery of natural gas from a single well.

According to Engelder & Lash (2008), the Marcellus Shale has water saturation of around 1,100 mg of H20 per m^3 of natural gas, which must be removed in order to transport it through pipelines, which typically have specifications so that the water content in the gas is restricted to 664 - 112 mg

H2O/ m^3 of gas. There are various estimates of the concentration of natural gas in the Marcellus in the literature, with Haper (2008) estimating 30 TCF in reserves and around 10 per cent of this in recoverable reserves. Compared to the average annual consumption of natural gas in the USA, in the order of 21 TCF, the Marcellus field is classified as a "super giant" by Lee et.al. (2011).

The Marcellus formation is currently producing natural gas, recovered from vertical and horizontal wells, which are accompanied by the hydraulic billing technique and whose intersection varies between 600 and 1500 metres laterally. According to estimates by Engelder et al.(2008), horizontal wells have produced around 4 billion cubic feet (BCF) of natural gas in the formation based on the Estimated *Ultimate Recovery (*EUR) calculation, corresponding to three times the production of vertical wells. However, they have a cost of around 3.5 million USD, compared to 1 million USD in vertical wells (Engelder & Lash, 2008), Horizontal wells have a higher recovery rate and, in the Marcellus, are drilled in such a way as to intersect as many fractures as possible perpendicularly in order to obtain greater productivity from the well (billing pattern in the Marcellus): east-northeast direction, then drill north-northwest, intersecting the set of fractures perpendicularly).

The application of the hydraulic fracturing technique results in the generation of micro-fractures along the lithology, which allow and favour the passage of fluid into the well. These artificially generated fractures tend to follow the pattern of the billing already existing in the rock. In the case of the Marcellus Shale, microfractures are generated across and parallel to the J1 set of fractures, stimulating induced permeability and, consequently, hydrocarbon recovery. However, this induced turnover process must be controlled in order to prevent fracture extensions and consequent leakage; in the case of the Marcellus, the formation is overlain by a carbonate sequence called Onondaga, which helps to prevent the continuation of these fractures, acting as a sealing rock for this system (Lee et.al.2008). Despite the pattern of exponential decline in production, which is generally reduced to a fifth of the amount initially produced over a period of five years, the Marcellus has maintained considerable production values; mainly due to the implementation of recovery techniques such as acid and/or carbon dioxide injection, which encourage the recovery of hydrocarbons at relatively low costs.

Among the main environmental consequences associated with the exploitation of this resource are: the expansion of road traffic in the region, since the transport of drilling equipment fluids is essentially carried out by lorries, which interfere with local infrastructure, generate potential contamination of water bodies in the region and the emission of pollutants from these vehicles, as well as indirectly interfering with local fauna; excessive water consumption, which has been estimated at between 6.1 to 18.7 million gallons per day (MGD), corresponding to around 1% of the daily consumption in the state of Pennsylvania. Although not alarming, excessive water exploitation can be damaging during periods of drought in the region. The main environmental issue associated with the exploitation of unconventional natural gas is related to the disposal of the water used in the process, where around 30 to 70% of the water used in fracturing returns to the surface, in addition to the water removed from the geological formation being recovered. The water resulting from this process has its composition altered due to the addition of chemical compounds, as well as the consequent enrichment in total dissolved *solids* (TDS), in the concentration of metals and radioactive elements from organic-rich shales. The concentration of contaminants is controlled by regulation, mediated by the *Environmental Protection Agency (*EPA), which sets standards and limits on the concentration allowed per component (Lee et al. 2011).

Several alternatives are already on the market for this water disposal. Firstly, the water used can be treated according to EPA standards for drinking water and made suitable for human consumption, however, due to the need to remove all types of contaminants present, the process becomes costly. Another option would be to inject this water into reservoirs in which confining layers would act to prevent contamination of aquifers from this fluid (Soeder et.al., 2009). Finally, the water can be treated to be reused in the drilling and fracturing process itself, which represents a considerably less intensive and costly treatment than the treatment to make it drinkable; and has been extensively researched and used by various operating companies in the sector. In this case, additional water treatment plants should be installed in the region, as municipal plants are generally not properly equipped to treat the water used in this process.

According to Harper (2008), natural gas production in the Marcellus in the state of Pennsylvania from 2009 to 2011 has averaged 84,000 thousand cubic *feet* (TCF) of gas per day, a production rate that, according to the author, can be maintained for around 20 years. However, production figures for western Pennsylvania and eastern Ohio are higher than the average.

- .1.2 BARNETT SHALE

The Barnett geological formation is part of the North American Fort Worth sedimentary basin, currently located in the north-central part of the state of Texas, covering a total area of approximately 28,000 miles^2 (approximately 45,000 km^2). The formation varies in thickness and depth, with outcrops in the south-eastern part of the basin and subsurface dipping in the Texas-Oklahoma direction, where it is deeper and thicker. The geographical limits of the formation are represented by the Ouachita faults to the east, the Muenster and Red River arcs to the north, and the Eastern Platform and Concho arc to the west. The portion that generates unconventional natural gas (shale gas) corresponds to around a third of the total length of the formation (approximately 9,000 miles2 or 23,000 km2), and the production of this resource is concentrated in the north of the basin, where the Barnett formation is thicker (Montgomery et al., 2005).

The Fort Worth basin has been explored since 1900 for conventional hydrocarbon production. Since 1998, the recognition of geological formations corresponding to continuous accumulations of non-associated natural gas, attributed to the Barnett Shale, has promoted the exploration of unconventional natural gas from this formation (Loucks & Ruppel, 2007). The basin dates back to the upper Palaeozoic and its formation is associated with the Ouachita orogeny, which resulted from the convergence of the Laurussia and Gondwana megacontinents. The basin is characterised as foreland and was located in the southern portion, on the edge of Laurussia, as shown in figure 32. Currently, the basin is classified as shallow, asymmetrical and with a north-south structural axis, parallel to the Ouachita lineament. The basin is structurally deeper in its north-eastern portion and gently shallower in its western and south-western directions, where Precambrian and Palaeozoic rocks are exposed (Figure 33) (Loucks & Ruppel, 2007).

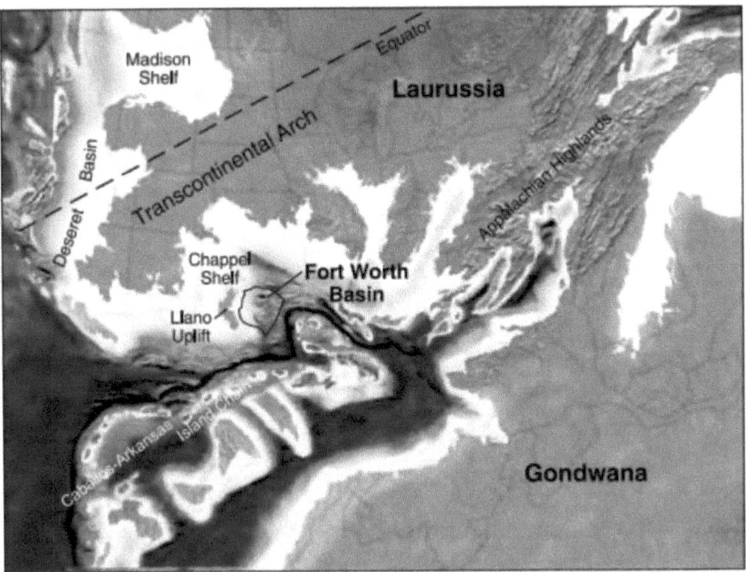

Figura 32 - Representation of the regional palaeogeography of the upper Mississippian (325 Ma), indicating the location and formational geological context of the Fort Worth sedimentary basin. Source: Loucks & Ruppel (2007).

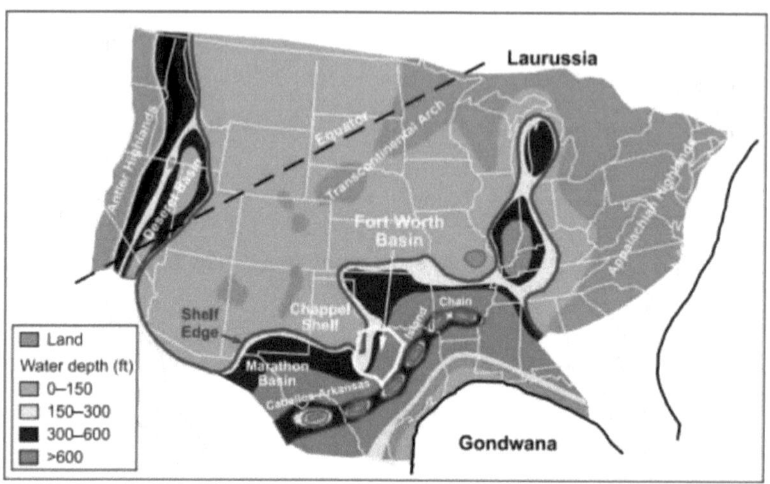

Figura 33 - Palaeogeographic map of the United States showing the depth variations present in the Fort Worth basin and surrounding areas, indicating more favourable zones for the occurrence of unconventional natural gas (greater depths). Source: Loucks & Ruppel (2007).

Based on the stratigraphy of the basin, it can be said that the Barnett Formation corresponds to a set of laminated shales, with a siliceous and carbonate composition, and carbonates. The predominant composition is represented by black shales rich in organic matter, fossiliferous, and dense, dark-coloured carbonates, also fossiliferous, both dating from the upper Mississippian (Lancaster et al., 1993). In the north-eastern portion of the Fort Worth basin, the Barnett Formation is informally divided into two members: upper and lower; a division based on the increased carbonate content (higher in the upper member). However, both members are intercalated with carbonate layers (calcites and dolomites).

According to Bruner & Smosna (2011), the Barnett Shale is quite variable in depth, starting at 50 feet, with a maximum depth of around 10,000 feet (approximately between 15 and 3000 metres). Its lithology is basically made up of limestones and black shales, which vary compositionally between siliceous and carbonaceous. Gamma-ray analyses of the shales indicate values ranging from 150 to 200° API and sample densities of around 2.40 and 2.50 g/cm^3 . The porosity of the formation is around 6% and water saturation between 20 and 30%. Gas saturation was measured at between 70 and 80 per cent, which is predominantly stored in the interstices of the pores and in microfractures. The pressure in the formation varies between 3000 and 4000 psi. Total organic carbon (TOC) values vary between 2 and 12 per cent throughout the formation, with the highest values corresponding to the lower member of the formation, with predominantly type II organic matter, but type III also present. Vitrinite reflectance values (Ro) of around 1.2%.

The Barnett Formation is delimited in the Fort Worth basin by orogenic zones: the Ouachita Thrust to the east, the Muesnter and Red River arcs to the north, and the Eastern Platform and Concho Arch to the west (Montgomery et al. 2005). The tectonic activity in the basin has resulted in the presence of several unconformities in its stratigraphic section (Figure 34). Among the shale gas producing fields in the Barnett formation (shale plays), the main one is the Newark East field, located in Denton, Wise and Tarrant counties. This producing field covers an area of around 1300 km^2 , and has around 2400 producing wells and an estimated potential of 2.7 trillion cubic feet of gas (Tcf) in proven reserves (Durham, 2005; Montgomery and et al., 2005).

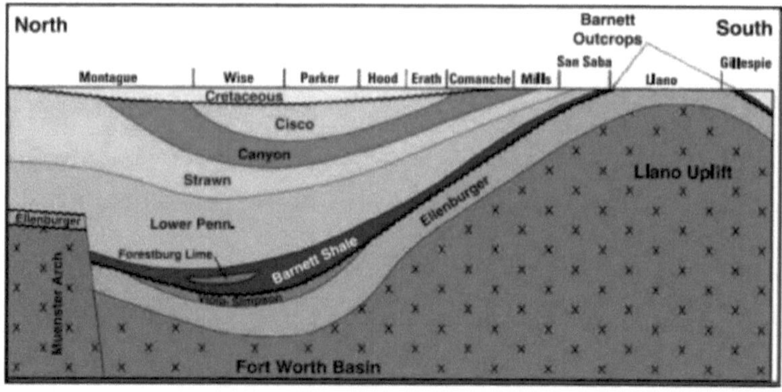

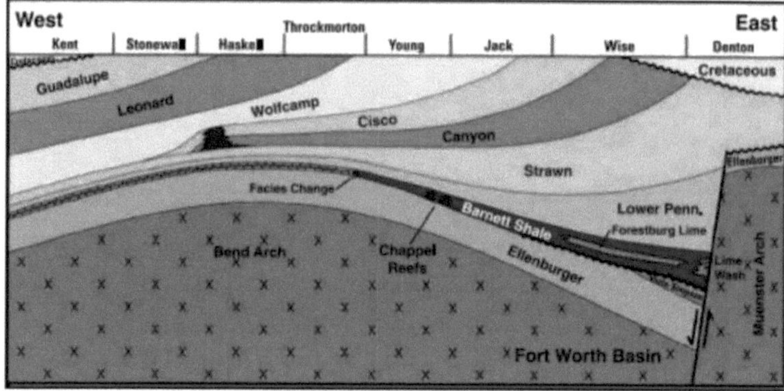

Figura 34 - **North-south and east-west stratigraphic sections of the Fort Worth sedimentary basin, representing the structural positioning of the Barnett Formation along the basin.** Source: Bruner & Smosna (2011).

Based on lithostratigraphic analyses by Loucks & Ruppel (2007), the mineralogy of the Barnett Formation is made up of approximately 35 to 50% silica/quartz and less than 35% clay minerals. Detailed analyses indicate: 45% quartz, 27% illite and smectite, 8% calcite and dolomite, 7% feldspars, 5% organic matter, 5% pyrite, and 3% pyrite and trace elements (Bowker, 2002 apud Bruner & Smosna, 2011).

The depositional environment of the Barnett was determined by Loucks & Ruppel (2007) as a marine platform, with the presence of anaerobic conditions (deposition of black shale and organic matter and pyrite) and aerobic conditions (carbonates). The production of hydrocarbons from this formation is predominantly thermogenic natural gas (around 90 per cent). Carbon dioxide

(CO_2) accounts for approximately 2% of the total volume of gas produced; nitrogen and sulphur occur in trace fractions.

■ .1.3. EAGLE FORD SHALE

Eagle Ford corresponds to a hydrocarbon-generating geological formation dating from the Upper Cretaceous and associated with the marine-continental shelf depositional environment. Its composition is predominantly represented by carbonaceous shales rich in organic matter (40-90%), clay minerals (15-30%), and quartz/silica (15-20%). Its total organic carbon (TOC) content varies between 2 and 12%, thermal maturity (Ro) between 0.45 and 1.4%, API between 28 and 62°, porosity between 8 and 12%, and pressure gradient (psi/ft) between 0.5 and 0.8, and the thickness of the generating portion varies between 45 and 450 feet (12 to 137 metres). A

The depth at which the formation is found varies greatly, generally exceeding 4,000 feet and reaching up to 14,000 feet (around 1,200 to 4,200 metres) (EIA, 2014).

According to Za Za Energy (2013 apud EIA, 2014), the Eagle Ford formation is divided into two different members: an upper unit, characterised by intercalations of carbonate rocks and shales whose deposition is associated with transgressive regimes (low sea intervals); and a lower unit, predominantly made up of black shales deposited in a transgressive regime (high sea intervals). Eagle Ford is overlain by limestones of the Buda Formation and overlain by the Austin Chalk. The upper member has a higher carbonate/clay ratio than the lower member (Parra et al., 2013).

The Eagle Ford geological formation covers an area of approximately 20,000 square miles (52,000 km^2) in the southern and southeastern portions of Texas. The formation is originally known as a hydrocarbon source rock in the Austin Chalk petroleum system, which produces oil and natural gas, and dips towards the Gulf of Mexico. The formation is essentially made up of black shales and carbonates, with the predominant carbonate content accounting for around 70% of the formation; a composition which, due to the high carbonate content and low clay content, gives the unit a more fragile and rutile behaviour, favouring the implementation of the hydraulic billing technique (Matsutsuyu, 2011).

The Eagle Ford geological formation is part of the Maverick sedimentary basin, currently located in the state of Texas, and is the thickest formation in the basin. The extent of the Eagle Ford formation and variations in its thickness, depth and stratigraphy throughout the basin are mainly attributed to regional tectonic aspects, which involve the Maverick basin, the San Marcos arc, the margin of the Stuart City and Sligo continental shelves, and the East Texas basin, represented in figure 35 (Hentz & Ruppel, 2011).

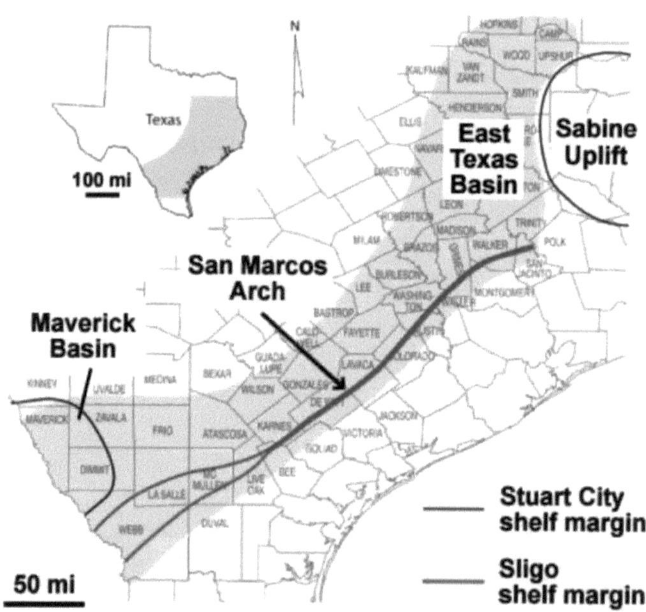

Figure 35 - Representative map of the location and main tectonic-structural features related to the Eagle Ford formation, southern portion of the state of Texas. Source: EIA (2014).

From 2008 to 2012, Eagle Ford became one of the most actively explored shale gas fields in the United States, with vertical and horizontal oil wells producing up to 1,000 barrels of oil per day and more than 200 exploratory rigs.

Exploration and production in the Eagle Ford intensified in 2008, inspired by the productivity and economic return from the Barnett Formation in the Fort Worth Basin, also in Texas (EIA, 2014).

9.2. Comparative Analysis

Many hydrocarbon exploration and production programmes are based on analogies with proven producing oil fields. As an example, the Barnett field, in the Fort Worth sedimentary basin, discovered in 1981, had its exploration encouraged due to comparisons with the Devonian shale of the Appalachian (Marcellus) basin, which has been proven to produce oil (Steward, 2007 apud Bruner & Smosna, 2011). In this context, studies to review the literature and compile information on the geological characterisation of these shale gas producing fields, especially comparative parameters such as reservoir thickness, organic carbon content, thermal maturity, gas volume and adsorbed gas fraction, are essential in determining the optimum conditions for the occurrence of unconventional natural gas in different locations. In this context, this research project proposes a comparative analysis of the main fields producing unconventional natural gas in the United States (the largest producer and a pioneer in the production of this resource), in order to determine, by analogy, the potential for this resource to occur in Brazil; more specifically in the Irati and Ponta Grossa formations of the Paraná Basin.

From a geological point of view, the Marcellus, Barnett and Eagle Ford producing fields are relatively similar: the formations are part of a context of accumulation in foreland sedimentary basins, along converging tectonic plate margins delimited by structural boundaries of faults and fractures resulting from orogenies, which contributed to the increase in the thickness of the sedimentary packages.

Their lithostratigraphic compositions also coincide: black shales rich in organic matter and carbonates; they are also divided into members, based on variations in the predominance of these lithologies and the clay/silica ratio. In terms of geochemistry, the Eagle Ford formation shows significant variations, while the Barnett and Marcellus formations are more similar: despite small differences in geochemical composition, both formations generally show similar gamma-ray analysis values: in the order of 150 to 200 API, and are found at depths of between 500 and 1000 feet (150 to 300 metres). In terms of basic reservoir characteristics, the formations are quite similar, according to analyses by Bruner & Smosna, (2011):

- Porosity: approximately 6%;
- Water saturation: between 20 and 30 per cent, contact with clay minerals, no free water;
- Gas saturation: between 70 and 80 per cent, present in interstitial pores, microfractures, and adsorbed to solid organic matter and kerogen;
- Adsorbed gas: minimum values of 20 to 25 per cent and maximum values of 40 to 60 per cent;
- Internal reservoir pressure: predominantly stages of normal pressure to overpressure (0.46 to 0.52 psi/ft); although Marcellus presents portions of low pressure (0.10 to 0.35 psi/ft);
- Forming pressure: from 3,000 to 4,000 psi;
- Drilling depth: 4,000 to 8,500 ft in Barnett (1,200 to 2,500 m) and 2,000 to 10,000 ft in Marcellus (600 to 3,000);

The Eagle Ford formation, on the other hand, has lower gamma ray values (between 28 and 62°API), and higher porosity values (between 8 and 12 per cent) and pressure gradients (between 0.5 and 0.8 psi/ft, and an average of 0.65 psi/ft). In addition, drilling reaches depths of around 11,500 ft (around 3,500 m) (EIA, 2014).

In terms of hydrocarbon generation, the formations differ: Eagle Ford has an estimated natural gas potential of 278 trillion cubic feet (Tcf), while Barnett has around 390 Tcf and Marcellus 500 Tcf. The total organic carbon content (TOC%) in these formations varies between 2 and 12% (predominantly 2 to 6% in Barnett, 2 to 10% in Marcellus, and 2 to 12% in Eagle Ford). The formations described are essentially made up of type II and III kerogen, with the Barnett and Eagle Ford fields having a higher fraction of type II kerogen (more favourable for natural gas generation) than the Marcellus. The thermal maturity of the sedimentary basins in which these formations are located is attributed to post-depositional burial, which is relatively uniform throughout the basins and, consequently, more homogeneous in the formations; showing only local variations in the vicinity of fault and fracture zones and hydrothermal influence. Vitrinite reflectance values (Ro) are in the order of 1.4 in Eagle Ford, 1.2 in Barnett and 1.6 in Marcellus; a difference related to the speed of heating and the rate of maturation (higher in Marcellus due to palaeoenvironmental conditions) (Bruner & Smosna, 2011).

Based on the data presented by Bruner & Smosna (2011) and Norton Rose Fulbright (2013), the following descriptive-comparative table was developed, which summarises the main parameters essential to the occurrence of unconventional natural gas (shale gas) and compares these parameters between the main fields producing this resource up to 2014: Marcellus, Barnett and Eagle Ford (Table 8).

Table 8 - Comparative analysis of the Marcellus, Barnett and Eagle Ford fields, proven generators and producers of unconventional natural gas.

Comparative Parameters	Marcellus	Barnett	Eagle Ford
Geological Age	Middle Devonian	Upper Mississippian	Upper Cretaceous
Lithology	Carbonaceous, highly radioactive black shales with concretions carbonates; shales enriched in silica and organic matter.	Dense, siliceous, oil-bearing, fossiliferous black shales rich in organic matter; crystalline, oil-bearing and fossiliferous carbonates; black shales highly enriched in silica (more rutile).	Carbonaceous shales rich in organic matter; petroleum and fossiliferous carbonates also enriched in organic matter.
Mineralogical composition Predominant (% volume)	Silica (10-20%); clay minerals (35%); carbonates (50%)	Silica (35-50%); clay minerals (<35%); carbonates (20%)	Silica (15-20%); clay minerals (15-30%); carbonates (40-90%)
Mineralogy (% volume)	27-31% quartz, 9-34% illite, 1-7%	30-50% quartz, 27% illite and traces	15-20% quartz, 15-30% clay

	clay mineral complex, 0-4% chlorite, 3-48% calcite, 0-10% dolomite, 0-4% sodium feldspar, 5-13% pyrite, and 0-6% gypsum.	of smectite, 8-19% calcite and dolomite, 7% feldspar, 5% organic matter, 5% pyrite, 3% siderite, and traces of phosphate and copper.	minerals, 40-90% calcite and dolomite + organic matter + pyrite.
Area (km)[2]	60.000	13.000	12.140
Thickness (m)	15-200 (typically 15-80)	15-300 (typically 45-180).	15-150 (typically 70)
Depth (m)	600-3.000	1.200-2.500	1.200-4.000
Type of Kerogen Predominate	II and III (predominant type II)	II (only type III traits)	II and III
Ro (%)	1.60 (maximum 3.5)	1.2 (maximum 1.9)	1.4
TOC (%)	1-10	2-6	2-12
Tmax (°C)	470	465	335
Thermal Maturity	Predominantly supermature rocks (thermogenic gas). Associated with the depth of burial and thermal flow in fault zones.	Rocks that vary from immature to supermature. Associated with the depth of burial and hydrothermal activity along fault and fracture zones.	Rocks that vary from immature to supermature. Associated with the depth of burial and hydrothermal activity along fault and fracture zones.
Basin Classification Sedimentary	Antepaís, asymmetrical, delimited by orogenic belts, structural arches and Precambrian uplift.	Antepaís, asymmetrical, delimited by prominent orogenic belts, structural arches and Precambrian uplift.	Antepaís, asymmetrical, delimited by orogenic belts.
Depositional palaeoenvironment	Deep-water sedimentary basin, low circulation, stratified water column, reducing environment at the base, presence of algae and planktonic organic matter,	Deep-water sedimentary basin, low circulation, stratified water column, reducing environment at the base, presence of algae-planktonic organic matter, water depth of around 50 to 200	Sedimentary basin on a marine-continental platform, low circulation, stratified water column, reducing environment at the base, presence of

	water depth of around 300 to 300 metres.	metres.	organic-algae-planktonic matter , water depth of around 100 metres.
Reservoir characteristics (%)	φ= 6, k = micro-nanodarcies, Sw = 30 (no free water), Sg = 70 (gas allocated to interstitial pores and microfractures, and adsorbed to organic matter and kerogen: 20-80%); internal reservoir pressure: from 0.51 psi/ft (overpressure) to 0.10-0.35 psi/ft (low pressure)	φ = 3-6, k = micro-nanodarcies, Sw = 25 (no free water), Sg = 75 (gas allocated to interstitial pores and microfractures, and adsorbed to organic matter and kerogen: 20-60%); overpressure stage: 0.52psi/ft	φ = 8, k = micro-nanodarcies, Sw = 20 (no free water), Sg = 75 (gas allocated to interstitial pores and microfractures, and adsorbed to organic matter and kerogen: 20-80%)internal reservoir pressure: 0.5-0.8 psi/ft, average 0.65 psi/ft (overpressure)

CHAPTER 10

REGIONAL GEOLOGY OF THE PARANÁ BASIN

10.1. General Aspects

The Paraná Basin corresponds to a vast region of sedimentation in South America and is characterised as an intracratonic basin. Located in the central-eastern portion of the South American Platform, the basin comprises territorial portions in southern Brazil, eastern Paraguay, north-eastern Argentina and northern Uruguay (Figure 36), which total an area of approximately 1.4 million square kilometres: 1.1 million km^2 in Brazil, 100,000 km2 in Paraguay, Argentina and Uruguay. The Brazilian portion of the basin includes, in significant proportions, the states of São Paulo, Paraná, Santa Catarina and Rio Grande do Sul, and in smaller proportions, the states of Mato Grosso, Mato Grosso do Sul, Goiás and south-western Minas Gerais (Figure 37) (Zalán, et al. 1988, 1990a).

Figure 36 - Location map of the Paraná Basin on the South American Platform. Source: Zalán et al. (1990).

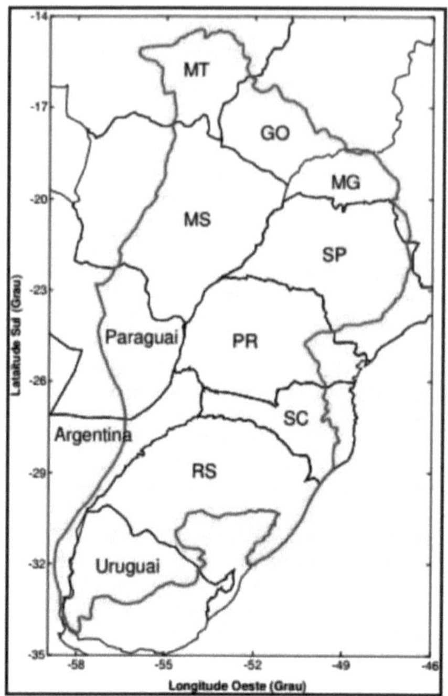

Figure 37 - Location map of the Paraná Basin (red outline) showing its boundaries in the Brazilian states (MT, MS, GO, MG, SP, PR, SC, RS) and neighbouring countries (Paraguay, Argentina and Uruguay). Source: Lima Gomes (2009).

Inserted in the paleocontinental geological context of Gondwana, the Paraná Basin's geological evolution is attributed to the Paleozoic and Mesozoic periods, presenting a magmatic and sedimentary stratigraphic record with a maximum total thickness of around 7,000 metres, whose spatial positioning coincides geographically with the structural depocentre of the syncline with the Paraná River channel, from which, according to Maack (1947), the basin was named. Its temporal position varies between the Neo-Ordovician and Neocretaceous (Milani, 1997). Its tectono-stratigraphic evolution encompasses the set and succession of processes that include continental distension, thermal subsidence and isostatic adjustments (Milani, 2004).

The basin's contour forms an erosional boundary along much of its perimeter, reaching a diameter of approximately 5,500 kilometres.

Morphologically, the Paraná Basin predominantly represents post-Palaeozoic events on the South American continent. To the east, the basin's flank, sculpted by erosion resulting from the marginal uplift of the South Atlantic rift, significantly exposes the crystalline basement; while to the west, the basin is delimited by the Asunción Arc, which results from lithospheric overloading due to the Andean belt and which segregates the Paraná Basin from areas previously belonging to it, such as the Paraguayan-Bolivian chaco. The north-east of the basin shows an original boundary, represented by a depositional palaeobord. The basin extends to the territories of Uruguay and Paraguay to the south-west (Milani, 2004).

Milani (1997) identified six large-scale stratigraphic units or supersequences in the basin: Rio Ivaí (Ordovician-Silurian), Paraná (Devonian), Gondwana I (Carboniferous-Eotriassic), Gondwana II (Meso to Neotriassic), Gondwana III (Neojurassic-Eocretaceous) and Bauru (Neocretaceous). According to the author, the first three supersequences represent transgressive-regressive cycles related to the oscillation of the relative sea level characteristic of the Palaeozoic; and the others correspond to Mesozoic continental sedimentary packages associated with igneous intrusions (Figure 38).

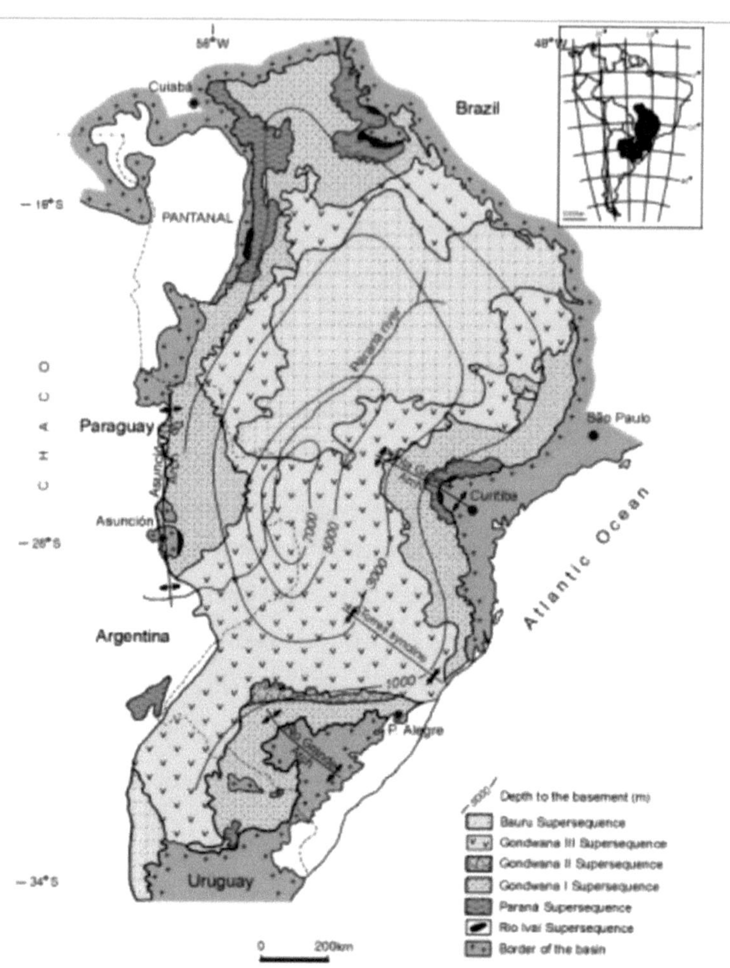

Figure 38 - Simplified geological map of the Paraná Basin; with structural outline (depth) of the crystalline basement. Source: Milani (2004).

The subject of the geology of the Paraná Basin has been studied historically and includes several hundred published works dealing with different aspects of this issue, making it one of the main themes in the Brazilian geoscientific bibliography (Milani et al. 2007). In this historical context, we highlight the work of White (1908), considered the "ground zero" of the stratigraphic characterisation of the basin, as well as the work of Sanford and Lange (1960), Northfleet et al. (1969), Schneider et al. (1974), Soares et al. (1978), Almeida (1980), Fulfaro et al.

(1980) and Zalán et al. (1990), referring mainly to the regional geology of the basin, plus the works linked to PETROBRAS, such as those by Milani (1997) and Milani et al. (2007), which were adopted as the main basis for preparing this chapter on regional geology.

Information on the main structural and stratigraphic features of the Brazilian portion of the Paraná Basin was obtained from a programme of drilling and geophysical surveys carried out by PETROBRAS with the aim of mapping and identifying sites suitable for the accumulation of hydrocarbons in the basin (Quintas et al. 1997). According to Yoshida and Gama Jr. (2006), from 1953 to 1974, a major systematic effort was made to search for oil and natural gas reserves in the Paraná Basin, which was mainly motivated by its proximity to the main consumer centre for these resources in Brazil (the south and southeast regions). During this period, 71 exploratory wells were drilled and different prospecting methods were applied, generally of an experimental nature, such as aeromagnetometry, gravimetry and reflection seismic. Even though these activities were limited to a strip of land corresponding to 30 per cent of the total length of the Paraná Basin and despite their unpromising results for conventional hydrocarbon accumulation, they made a significant contribution to building up a body of information on this basin.

10.1.1.TECTONOSTRUCTURAL FRAMEWORK

According to Milani (2004), the main problem in the study of the origin of the Paraná Basin is related to the limitations in visualising its deep framework, in the region known as the "central trough" of the basin, along the axis of the Paraná River. This is classified as a central rift, identified through regional geophysical data (gravimetry, magnetometry, reflection seismic, as well as lithostratigraphic analyses of borehole cores) and which corresponds to a prominent pre-Devonian depression (Figure 39) geographically delimited by the axis of the Paraná River drainage basin, and geologically by normal faults. According to the classification of Marques et al (2003), the central rift is made up of an Ordovician-Silurian domain internally compartmentalised into a series of structural highs and lows. The author also states that these pre-Devonian sedimentary successions are thicker than the others because they are conditioned by the central rift, which extends for around 600 km in a SW-NE direction (from Paraguay to the north-central part of the basin).

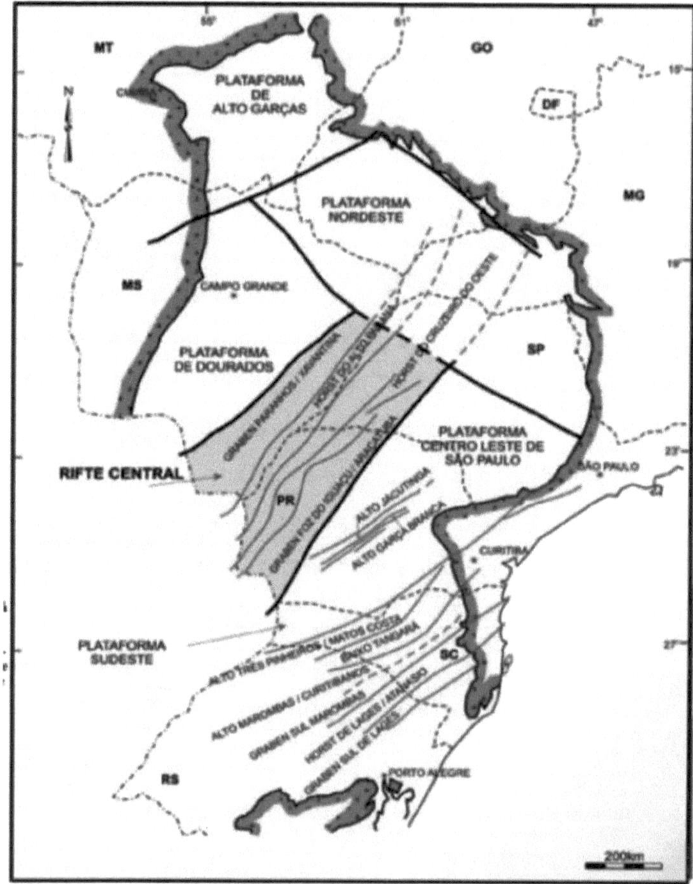

Figure 39- Map of the tectonic framework of the Paraná Basin.
Source: Marques et al. (2003) apud Milani (2004).

After the formation of the central rift, which is associated with stretching resulting from lithospheric heating and mechanical subsidence promoted by normal faults, there is a stage of thermal subsidence resulting from the dissipation of the heat accumulated previously (Marques et al, 2003), allowing the Devonian sedimentary package and the six major sequences of the Paraná Basin stratigraphic framework proposed by Milani (2007) to be accommodated.

10.2. TECTONO-SEDIMENTARY EVOLUTION

The Paraná Basin developed within the Gondwana supercontinent, initially as a wide gulf, open to the Panthalassa palaeo-ocean and later resulting in an intracratonic depression with episodes of orogeny (Zalán et al. 1990). According to Milani et al (2007), lithospheric flexure due to tectonic overload was the main mechanism of subsidence throughout the evolutionary history of the basin; an interpretation that differs from other authors such as Zálan et al. (1990), who associate the establishment of the syncline with the techno-magmatic phenomena of the Brasiliano Cycle and the consequent thermal contraction, or Fulfaro et al. (1982 apud Milani et al. 2007), who determine a set of aulacogenic troughs, orientated in a NW-SE direction, as precursors of cratonic sedimentation.

According to Milani (1997), based on analyses of the Precambrian substrate, "the basin would have been established in the form of elongated depressions in a NE-SW direction". Also according to Milani (1997) and Ramos et al. (1986 apud Milani et al. 2007), the reactivation of crustal discontinuities present in the basin's basement, which correspond to the Brasiliano archabasement and are located under the compressional field of the continent's edge as a result of the Neo-Ordovician orogeny, would have favoured the accommodation of its first cratonic unit: the Rio Ivaí supersequence, whose sedimentary package has its top marked by a Neossilurian discordance.

The Paraná supersequence, of Devonian age and characterised by its facies uniformity, would have accumulated with the resumption of subsidence in the basin. The corresponding sedimentary package has a varied thickness and a top demarcated by successive erosive events dating from the Neodevonian to the Middle Carboniferous. The authors identified a gap in the techno-sedimentary evolutionary history of the Paraná basin, represented by a hiatus of around 70 million years corresponding to the Mississippian period of geological time, which is absent from the basin record. This gap would have been justified by tectonic factors associated with the Hercynian Orogeny and/or the intermittent development of ice caps over the basin, in addition to the significant lowering of the sea level related to these glaciations (Milani et al. 2007). Accumulated after the Neodevonian discordance, the Gondwana I Supersequence was accumulated as the process of continental separation continued, with Gondwana migrating northwards and the consequent deposition of carboniferous sedimentation. Its sedimentary package corresponds to a complete transgressive-regressive cycle: invasion and retreat of Panthalassa over the interior of Gondwana. Its sedimentary cycle is also influenced by the melting of the polar ice caps present, characterised by mass flow and resedimentation. The conditions of maximum flooding date back to the Artinskian and the supersequence ends at the beginning of the Triassic with continental depositional systems. Milani et al (2007) state that the accumulation of Gondwana I is concomitant with the progressive closure of the Paraná basin, which initiates the intracratonic character of the basin as it is progressively trapped in the continental interior of Gondwana during the Mesozoic period.

The Triassic period in the Paraná Basin is characterised by a generalised distension of the Gondwana palaeocontinent in its southern portion, a geological context in which the Gondwana II superquence would have been deposited. This superquence is restricted to the southern portion of the basin (in the state of Rio Grande do Sul and Uruguay) and has sedimentation similar to that of graben-type basins. Large-scale erosive conditions throughout the Mesozoic, associated with aeolian abrasion within the Gondwana palaeocontinent, represent another stratigraphic gap in the basin, which is even more significant. The subsequent accumulation of dune fields from the Jurassic period onwards, succeeded by Eocretaceous magmatism resulting from the initial break-up of the palaeocontinent, makes up the sedimentary assemblage referring to the Gondwana III supersequence (Milani et al 2007).

The Eocretaceous period in the basin's geological record is represented by intense rifting and basaltic magmatism resulting from the break-up of the supercontinent Gondwana and the consequent evolution of the South Atlantic Ocean. This event is recorded in the Serra Geral formation, which represents basaltic flows and the end of extensive sedimentation within the supercontinent. Once the isostatic adjustment of the South American Platform was established in the face of the opening of the Atlantic Ocean, a depression was implanted over the igneous package, in which the sedimentary fill of the Bauru supersequence accumulated in the area corresponding to the Paraná Basin during the Neocretaceous. Fernandes & Coimbra (1996) do not attribute this Neocretaceous sequence to the Paraná Basin sedimentary record, associating it with a different basin: the Bauru Basin. Throughout the deposition of the Bauru supersequence, igneous activity was also present; represented by intrusive bodies often positioned on the northern edges of the basin (Figure 40) (Milani et al. 2007).

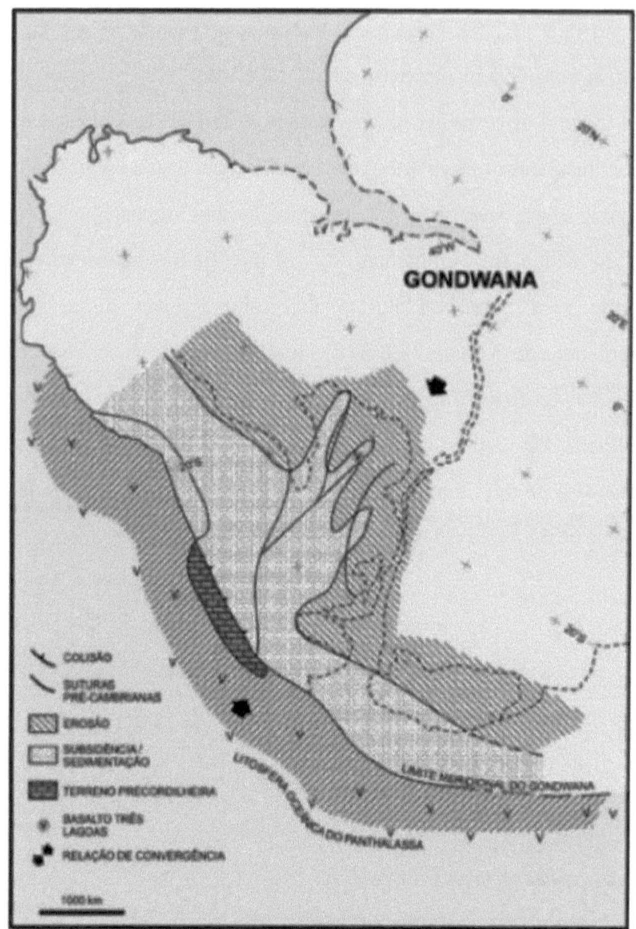

Figure 40 - Palaeotectonic model of the south-western margin of Gondwana during the Neo-Ordovician period (Ocloian Orogeny): initial subsidence phase of the Paraná Basin (delimited in dashed line). Source: Milani (2004).

10.3. LITHOSTRATIGRAPHIC FRAMEWORK

10.3.1. LITHOSTRATIGRAPHIC UNITS

10.3.1.1. Ivaí River Supersequence

Pre-Devonian sedimentary rocks were only conclusively dated and mapped in the Paraná Basin in the 1970s, initially on the northern flank of the basin through sampling and in the other portions through correlations. These rocks correspond to sedimentary and magmatic sequences that obey Milani's (1997) classification as the Rio Ivaí Supersequence, named after the river of the same name in the north-western part of the state of Paraná and whose drilled stratigraphic well reflects the lithology mentioned and also present in geoscientific literature under different names, such as: Ordovician-Silurian Sequence, Ordovician-Silurian Tectonosedimentary Sequence, or Rio Ivaí Group. This corresponds to the oldest geological unit identified in the Paraná Basin, directly above the syncline basement. The top of the sequence is demarcated by a regional-scale discordance that delimits the upper portion of the Rio Ivaí Sequence, represented by Eossilurian strata, and the basement of the next supersequence, of Eodevonian strata (Furnas Formation). The sedimentary package occurs in the interior of the basin from the state of Santa Catarina to the states of Mato Grosso and Goiás, showing a regional trend of thickening towards the west and extending into eastern Paraguay, where it is represented, by lithostratigraphic correlation, by the Caacupé and Itacurubi formations, where they reach thicknesses of approximately 1000 metres (Milani et al. 2007).

Milani et al (2007) describe the facies sequence of the Rio Ivaí Supersequence as representative of the first transgressive-regressive cycle of the Paraná Basin's cratonic sedimentation; essentially marine sedimentation whose transgressive portion extends from the base of the unit to the fossiliferous pelites of the Vila Maria Formation, indicative of the maximum flooding of this sedimentary cycle. From the interval of maximum flooding to the top of the unit, the regressive portion of the sedimentation manifests itself, which shows little expression in Brazilian territory due to the erosive activity associated with the "pre-Fumas Formation" discordance; but it is very present in Paraguay, corresponding to the fine, micaceous and fossiliferous sandstones of the Cariy Formation.

The Rio Ivaí package is made up of 3 lithostratigraphic units: Alto Garças, Iapó and Vila Maria (Assine et al.1994 apud Milani et al.2007).

10.3.1.1.1. Alto Garças Formation

Milani (1997) describes the Alto Garças Formation as a sedimentary package with a maximum thickness of around 300 metres and an essentially sandy composition with basal inclusions of quartz conglomerates or conglomeratic sandstones with cross-stratification. The lithology is represented by fine- to coarse-grained quartz sandstones, not very feldspathic, which may show silty-clay variations and a reddish colour in the upper portions of the packages. Structurally, the formation is characterised by the record of fluvial palaeocurrents, represented by recurrent cross-stratifications along the base of the unit, whose analysis of outcrops on the western flank of the basin indicates a south-westerly orientation (Milani et al. 2007).

According to Milani (1997), the Alto Garças Formation contains ichnofossils of the genus Skolithos, found in boreholes in the stratigraphic well with the same name as the formation. However, the fossil content of the unit is not significant and is not sufficient to determine its age.

10.3.1.1.2. Iapó Formation

According to Milani et al. (2007), the Iapó Formation is made up of diamictites with a silty-sandy matrix of varying colour. The presence of clasts of various types also characterises the formation. The diamictites of the Iapó Formation are widely distributed throughout the Paraná Basin, and are stratigraphically correlated from the northern flank to the south-central portion of the basin, in the state of Paraná. The contact between this formation and its successor (Alto Garças Formation) is observed abruptly in the sedimentary package, which indicates a discontinuity in the sedimentation record in the basin. The upper contact is with the fossiliferous pelitic deposits of the Vila Maria Formation.

10.3.1.1.3. Vila Maria Training

Initially described in the south-western portion of the state of Goiás by Faria (1982 apud Milani et al. 2007) as a cratonic sedimentary package that includes, at its base, diamictites followed by fossiliferous shales and sandstones, interspersed with siltstones, in its upper portion; representing the entire stratigraphic sequence overlying the Furnas Formation. Later described by Assine et al (1994 apud Milani et al.2007), the Vila Maria Formation was restricted to the fossiliferous pelitic package, with the diamictites subtracted from the formation because they corresponded to the previous formation (Iapó Formation). This pelitic package has gradual granulation to sandy fractions in its upper portion, and has an average thickness of a few tens of metres. The lithology is represented by reddish micaceous shales with a ferruginous appearance; locally it is dark grey with a high fossil content. In the upper part of the package, the lithology is represented by sandstones and siltstones whose structure corresponds to hummocky cross-stratifications (Milani et al. 2007). Faria (1982) describes the presence of contraction cracks in the upper interval of the formation, suggesting a possible subaerial exposure of the depositional surface.

In the context of fossil records, the Vila Maria Formation, together with its counterpart in Paraguay, the Vargas Pena Formation, represents the characteristic fossil content of the Silurian interval in the Paraná Basin. Marine megafossils such as graptolites, trilobites, brachiopods, gastropods, bivalves and ostracods are recorded in these units, as well as palynomorphs such as chitinozoans, myospores and acritarchs (Milani et al. 2007). Fossils of graptolites (of the genera Monograptus and Diplograptus), chitinozoans and myospores are the most important biostratigraphically and the most relevant in determining the age of the formation; which was placed in the Eossilurian stratigraphic interval by means of this association, confirmed by absolute dating (Rb-Sr) of the shales (435.9 ± 7.8 Ma) by Mizusaki et al (2002 apud Milani et al.2007), and equivalent to the age indicated by the graptolites and palynomorphs of the corresponding Paraguayan Vargas Pena Formation.

10.3.1.2. Paraná Supersequence

Devonian sedimentary packages, which occur in the Paraná Basin in the southern part of Brazil, have been the subject of geoscientific research for a long time; initially described by Derby (1878 apud Milani et al. 2007) and classified by Oliveira (1912 apud Milani et al. 2007) as the Devonian sequence of the basin under the following names: "Grés das Fumas", "Xistos de Ponta Grossa" and "Grés de Tibagi", and is also known as the Paraná Group, Devonian-Mississippian Tectonosedimentary Sequence, Devonian Sequence and Paraná Supersequence (Milani, 1997). The Devonian sedimentary package has a maximum thickness of approximately 800 metres and extends throughout the basin, where it is generally superimposed on the Ordovician-Silurian strata of the Ivaí Group, or locally on the basement of the syncline; as far as eastern Paraguay. The base of the package is delimited by a discordance, which shaped the deposition of the lower portion of the package in a regular and flattened manner, represented, according to Milani et al. (2007) by a sandy sedimentary blanket around 250 metres thick throughout its area of occurrence. The top of the package is also delimited by a discordance, which is regional in scale and dates from the late Devonian to the Carboniferous period, and which has subtracted records from this interval in the basin. The Paraná Group is divided into two units: a lower sandy package called the Furnas Formation and an overlying pelitic package called the Ponta Grossa Formation (Milani et al. 2007).

Stratigraphically, the Paraná Supersequence corresponds to the second transgressive-regressive cycle of the Paraná Basin; the Furnas Formation is the transgressive signature and the pelitic package of the Ponta Grossa Formation is the record of high sea conditions locally punctuated by sandy progradations; the most expressive of these progradations corresponds to the Tibagi Member. This lithological sequence represents the first great flood of the Devonian period (Praguian to Eo-Eemsian); the marine drowning of the basin persisted until the Frasian period (Milani et al. 2007). The authors also describe a second episode of marine expansion during the Mesodevonian (Eifelian to Givetian), which would have resulted in the connection between the Paraná and Parnaíba basins, a fact evidenced by palaeontological correlation of megafossils and palynomorphs.

10.3.1.2.1. Furnas Formation

Milani et al. (2007) describe the Fumas Formation as "a succession of white, medium to thick, kaolinitic quartz sandstones exhibiting cross-stratification of various kinds". They also describe the presence of conglomeratic beds, around 1 metre thick, along the base of the package. In the middle portion of the formation, medium-grained sandstones dominate, interspersed with thin sequences of siltstones and muscovite shales, representing a highly stratified interval of the formation and structurally characterised by herringbone cross-stratifications (Assine, 1996 apud Milani et al. 2007). In the upper part of the unit, medium- to coarse-grained sandstones are more present, also locally intercalated with very fine sandstones with hummocky cross-stratification; and progressively more clayey towards the top, representing a gradual transition to the overlying pelitic formation: the Ponta Grossa Formation (Milani et al. 2007).

According to Assine (1996), the sandstones of the Furnas Formation represent the end of a cycle of rich fossiliferous content in the basin, a fact that favours his hypothesis about shallow platform accumulation. Due to this scarcity of fossils, the lower part of the unit has not been consensually dated; it is possibly attributed to the end of the Silurian. The presence of pelites in the upper part of the unit, which contain continental plants and myospores from the Psilophytales group, changes the problematic dating picture, placing the top of the Furnas Formation in the upper Lochkovian (Gerrienne et al.2001 apud Milani et al.2007). The association of this dating with the gradual transition between the Furnas and Ponta Grossa formations confirms the attribution of the Paraná Supersequence, entirely or almost entirely, to the Devonian period.

10.3.1.2.2. Ponta Grossa Formation

The Ponta Grossa Formation is represented by Devonian pelitic formations that are around 600 metres deep in the subsurface and 300 metres deep in outcrops; it extends along the entire Brazilian portion of the Paraná Basin and extends into Uruguay, where it is called the Durazno Group (Milani et al. 2007). The formation was first described in the state of Paraná and is subdivided into 3 stratigraphic members: Jaguariaíva, Tibagi and São Domingos (Lange & Petri, 1967 apud Milani et al.2007). The lower member, Jaguariaíva, corresponds to the flooding over the transitional systems of the upper portion of the overlying formation, Furnas Formation, and is represented by shales, around 100 metres thick, with the presence of stratified sandstone lenses that show reworking by waves. In the upper portion of the package (in the upper 20 metres), there are laminated black and carbonaceous shales, which represent potential hydrocarbon generators in the central portion of the basin, as well as representing an important milestone in the stratigraphic correlation (Milani et al. 2007). The Tibagi member is of sand-silt composition and corresponds to the intermediate portion of the sedimentary package; it represents a regressive context of progradation of deltaic systems originating from the sandy sedimentary input from the north-eastern edge of the basin (Andrade & Camarço, 1982 apud Milani et al.2007). The São Domingos member is dominantly represented by shales and corresponds to a large-scale flood that ends the Devonian record of the syncline (Milani et al. 2007).

The biostratigraphy of the Ponta Grossa Formation was first described by Clarke (1913 apud Milani et al.2007), an author considered a milestone in the classification of macrofossils in the unit; followed by Lange & Sommer (1954 apud Milani et al.2007), who synthesised the knowledge about the Devonian palaeofauna and palaeoflora in the basin. The fossiliferous content of the palaeofauna is predominantly represented by marine invertebrates such as brachiopods, trilobites, bivalves, gastropods, annelids and echinoderms. The palaeoflora is represented by Psilophytales, Lepidodendrales, Lycopsidales and Hyniales, which show the evolutionary advance of the Devonian period. The formation has been precisely dated using palynostratigraphic zonation based on chitinozoa and myospores, which attributed the deposition of the Ponta Grossa Formation to the Praguian to non-terminal Neofrasian periods (Milani et al. 2007).

10.3.1.3. Gondwana I Supersequence

The Gondwana I Supersequence corresponds to the largest sedimentary volume in the Paraná Basin, occurring almost continuously along the perimeter of the syncline and reaching a maximum thickness of around 2,500 metres. Its sedimentary package is predominantly Palaeozoic, and has been the focus of studies in the basin due to the occurrence of deposits of attributed economic value, such as coal beds and uranium deposits (Rio Bonito Formation), and hydrocarbon source rocks and reservoirs. Initially described by Derby (1878) and White (1908), the supersequence is also known as the Pennsylvanian-Permian Tectonosedimentary Sequence or the Carboniferous-Eotriassic Sequence.

The unit encompasses a wide variety of depositional conditions within an eocarboniferous geological context, ranging from glacial influence to arid continental dune field conditions, indicating the arrival of the Mesozoic. From a lithostratigraphic perspective, this supersequence is divided into the following sedimentary packages: Itararé Group (Lago Azul, Campo Mourão and Taciba formations) and Aquidauana Formation, Guatá Group (Rio Bonito and Palermo formations), Passa Dois Group (Irati, Serra Alta, Teresina and Rio do Rastro formations) and Pirambóia and Sanga do Cabral formations.

During the Eocarboniferous period, the south-western portion of the Gondwana supercontinent was exposed to high latitudes and was consequently subjected to extensive continental glaciation, which inhibited sedimentary deposition and was reflected in the absence of records in the interval between the Neodevonian strata of the Ponta Grossa Formation and the Neocarboniferous strata of the Itararé Group and Aquidauana Formation. Sedimentation resumed as the palaeocontinent progressively migrated northwards, thus moving the focus of the glaciation away (Milani et al. 2007). These glaciogenic deposits of the Paraná Basin are represented by massive or stratified diamictites, with the presence of pebbles and blocks of different lithology, showing sedimentary transport by glaciers; and by massive sandy facies, graded or with unidirectional undulations corresponding to a context of turbiditic sedimentation. The unit's pelitic packages were associated with the process of settling when massive, or turbiditic sedimentation when laminated. Other glaciogenic lithotypes such as rhythmites (varvites) and tillites (locally in the Itararé Group) are also present, although on a smaller scale (Rocha- Campos, 1967 apud Milani et al.2007). The glaciogenic section of the Gondwana I Supersequence is most present in the north-central portion of the basin, where it reaches its maximum thickness of around 1,300 metres (Milani et al. 2007).

Following the deglaciation phenomenon, the consequent rise in relative sea level was recorded by Lavina & Lopes (1987 apud Milani et al.2007) as the "Permian transgression" of the basin; in the form of strata that are sometimes sandy and sometimes pelitic and that follow the glaciogenic package. These have been subdivided into 3 members, according to their lithological predominance: Triunfo, Paraguaçu and Siderópolis (Schneider et.al.1974 apud Milani et.al.2007). The first member shows evidence of tidal reworking of the deltaic lobes; the second shows marine influence, represented by pelitic packages of considerable thickness; while the third member is characterised by lagoon restriction and the development of peat bogs, which gave rise to the coal deposits of the Siderópolis Member (Milani et al.2007). The sedimentary package associated with the melting of the gondwanic cap is approximately 1,500 metres thick, is attributed to the Itararé Group, and is represented by the Aquidauana Formation in the northern part of the basin; it is lithologically made up of diamictites intercalated with sandstones, characterised by the presence of both glacioterrestrial and glaciomarine elements. The sedimentation pattern of the Itararé Group, from north to south, is succeeded by onlap sedimentation from south to north, which characterises the deposition of the Guatá Group and the maximum palaeobathymetric conditions are recorded in the Palermo Formation (early Neopermian). The overlying sedimentary section is regressive in nature and is attributed to the Passa Dois Group, evidenced by the present Eotriassic eolian deposits (Serra do Cabral and Pirambóia Formations) (Milani & Ramos, 1998).

The Rio Bonito Formation is made up of marine levels, generally represented by tempestites, or by large pelitic packages (Paraguaçu Member). The upper part of the package comprises fine-grained sandstones, siltstones and carbonaceous siltstones interspersed with coal strata (more concentrated in the southern part of the basin).

The Palermo Formation is lithologically made up of siltstones and sandy siltstones, yellowish-grey in colour, characterised by the presence of bioturbations. Locally, there are also fine sandstones with hummocky stratification, arranged in a lenticular form along the formation; and dark grey shales, which are correlated to the maximum flooding of the Gondwana I Supersequence. Based on biostratigraphic analyses and palynological zonation, the formation is temporally positioned in the Artinskian; and the group to which it belongs, the Guatá Group, between the Sakmarian and Artinskian (Souza, 2006).

The Irati Formation corresponds to accumulations of carbonates and evaporites in the northern part of the Paraná Basin, and bituminous shales in the southern part; these were deposited in a hypersaline environmental context, developed in the interior of the basin due to an event restricting water circulation between the Panthalassa Palaeocene and the syncline (Milani et al. 2007). The black shales of this formation stand out for their high level of organic content, reaching considerable values (around 23%), classifying them as potential hydrocarbon-generating rocks in the basin. The fossiliferous content of the formation is unique, with the vertebrate palaeofauna being represented by reptiles of the Mesosaurus and Stereosternum genera, which have suggested, through palaeontological correlations with South Africa, the theory of continental drift since the 1920s. Through this biostratigraphic correlation and dating of zircons found in volcanic ash, the geochronology of the Irati Formation is placed in the Neo-Artinskian (Santos et al. 2006).

The overlying formation, the Serra Alta Formation, corresponds to a package of thinly laminated dark grey shales, which represent a low-energy depositional context and consequent clay deposition associated with the "drowning of the Irati Gulf". This pelitic sequence is equivalent to the last important marine incursion in the Paraná Basin (Milani et al. 2007).

The Passa Dois Group, of post-Serra Alta sedimentation, developed in a large-scale regressive context, where continental systems became more involved in sedimentary accumulation and resulted in the progressive silting up of the basin.

In this context, predominantly pelitic deposits with traces of tidal action (Teresina Formation) are succeeded by a progradational system of red beds and deltaic lobes, lacustrine pelites, fluvial deposits and eolian sandstones (Rio do Rastro Formation), dating from the Upper Permian to the beginning of the Triassic and widely distributed from the northwest to the southeast of the basin (Lavina, 1988 apud Milani et al. 2007). The Teresina Formation has an equivalent in the portion of the basin referring to the state of São Paulo, the Corumbataí Formation, both of which correspond to a broad tidal system. The overlying Rio do Rastro Formation has a lateral equivalent, the Sanga do Cabral Formation, which represents the "advance of continental systems over the lacustrine remnants of the Passa Dois Group". The Sanga do Cabral Formation is analogous to the Pirambóia Formation, located in the northern part of the basin, and is equivalent to fluvial and eolian deposits (Milani et al. 2007).

In general, the Gondwana I Supersequence corresponds to a complete transgressive-regressive cycle in the Paraná Basin; its sedimentation was initiated by a Pennsylvanian glacial package at the base, followed by records of maximum marine drowning in the Artinskian (Palermo Formation) and closed by the deposition of continental deposits that would fill the syncline at the beginning of the Mesozoic period (Milani et al. 2007).

10.3.1.4. Gondwana II Supersequence

The Gondwana II Supersequence, described by Milani (1997), corresponds to a Triassic system represented by a fossiliferous pelitic sedimentary package called the Santa Maria Formation, limited to the south of the Paraná Basin (Rio Grande do Sul state) and without equivalence with other portions of the basin. This pelitic sequence is part of a thick, dominantly sandy and aphosiliferous section, analogous to the Permian interval between the Sanga do Cabral and Botucatu formations (Neojurassic-Eocretaceous). The lithology of the Santa Maria Formation is basically made up of red beds. Its fossiliferous content consists mainly of vertebrates (reptiles from the Dicynodontia, Cynodontia, Pseudosuchia,

Rhynnchocephalia and Saurischia groups), as well as fossil plants and invertebrates (Crustacea and Insecta), on the basis of which the formation has been chronologically positioned between the Mesotriassic and Neotriassic (Milani et al.2007).

The sedimentary package referring to the Santa Maria Formation shows an abrupt basal contact with the sandstones of the preceding formation. Minali et al.(2007) interpret this contact as "a rapid lacustrine transgression", indicating an "episode of accelerated subsidence of the substrate and development of a hungry basin", which would later be filled by progradational sandy inputs, Based on this analysis, the formation can be briefly described as a set of lacustrine pelites and sandstones.

sandy progrades, which is also covered, by abrupt contact, by sandstones of the overlying and equivalent Tacuarembó formations in Uruguay and Botucatu in Rio Grande do Sul. In their interpretation, the authors associate this Meso-Neotriassic subsidence in the Paraná Basin with the development of asymmetrical distensive grabens, which would have accommodated fluvial drainage over the flexural ramp of the system, accumulating pelites in the most subsiding portions. In other words, the Gondwana II Supersequence represents lacustrine-fluvial sedimentation resulting from a subsidence event in the Triassic grabens located in the southern Brazilian portion of the Paraná Basin (Milani et al. 2007).

10.3.1.5. Gondwana III supersequence

The Gondwana III Supersequence, also known as the Jurassic-Eotriassic Sequence (Milani et al. 1994), is widely distributed along the Paraná Basin and is made up of a continental sedimentary package represented predominantly by eolian facies (Botucatu Formation) and magmatites (Serra Geral Formation).

The Botucatu Formation is lithologically made up of medium- to fine-grained sandstones, well sorted (high degree of sphericity) and pink in colour with a matte appearance; structurally characterised by tangential cross-stratification.

Faciologically, the formation corresponds to the so-called "Botucatu desert", recognised through aeolian reworking of fluvial deposits underlying the dune field (Milani et al. 2007). Also present in this sedimentary package are lenticular fractions of medium to coarse sandstones associated with conglomeratic sandstones, representative of torrential episodes in an alluvial-fluvial context; and sandy or silty clayey rhythmites characterised by their plane-parallel stratification, indicative of a lacustrine environment (Milani et al. 2007).

The Serra Geral Formation corresponds to a large magmatic province associated with the intense large-scale fissure volcanism that affected the South American cratonic basins during the Mesozoic. Milani et al. (2007) associate this igneous activity with tensional fields and endogenous phenomena that acted in the fragmentation of the Pangea palocontinent. These events are represented in the Paraná Basin by thick covers of lavas, networks of dykes and multiple levels of sills intruded into the Palaeozoic sedimentary strata. The formation is also widely distributed throughout the basin and has a remaining thickness of around 2000 metres. Lithologically, the formation is made up of toleitic basalts and basaltic andesites, as well as the presence, on a smaller scale, of rhyolites and rhyodacites. Based on geochemical analyses by Bellieni et al. (1984) and Mantovani et al. (1985), compositional variations are present throughout this igneous cover: in the northern portion of the basin, rocks enriched in TiO2 predominate, while those in the southern portion are poor in this component; a factor resulting from the different degree of crustal assimilation by this magmatic ascendancy. Dating carried out by Renne et al.(1992 apud Milani et al.2007), using the Ar/Ar technique, associated with stratigraphic correlations, indicates that the Serra Geral magmatism was an event temporally positioned between 137 and 126 Ma.

10.3.1.6. Bauru Supersequence

Described by Milani (1997) as a psammitic siliciclastic unit accumulated in a semi-arid to desert context, the Bauru Supersequence is widely distributed along the Paraná Basin, covering the states of Minas Gerais, São Paulo, Mato Grosso, Mato Grosso do Sul, Goiás and Paraná, as well as the northeastern portion of Paraguay; totalling an area of occurrence of around 370,000 km^2 with a thickness of around 300 metres. The formation's basal contact with the previous basaltic sequence occurs through an erosional discordance (unconformity). The lower part of this sedimentary package is represented by thin strata with a brecciated appearance and angular basalt clasts in an immature sandy matrix.

The lithostratigraphy of the Paraná Basin was defined by analysing boreholes, outcrop samples and interpreting geophysical, geochemical and petrographic data obtained from the state company Petróleo Brasileiro S.A. (Petrobrás), as shown in figure 41.

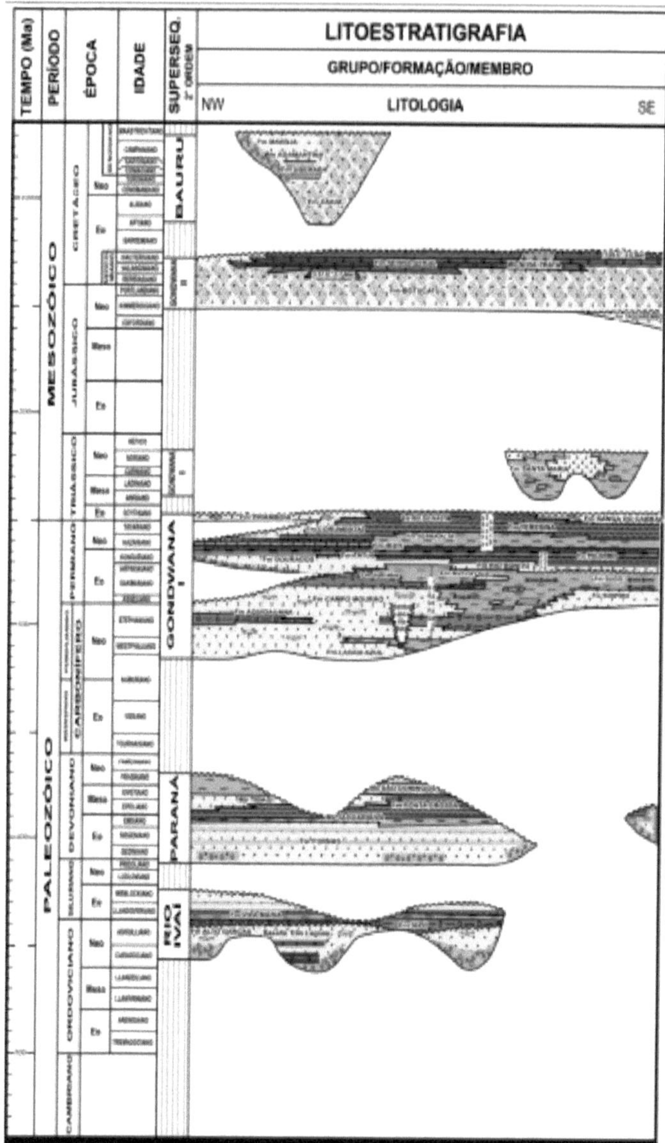

Figure 41 - Stratigraphic column of the Paraná basin.

Source: Milani (2004).

CHAPTER 11

POTENTIAL FOR HYDROCARBON GENERATION

The Paraná Basin has two main oil systems with a conventional configuration: the Ponta Grossa - Itararé formations and the Irati - Rio Bonito formations, with the pelitic formations (Ponta Grossa - Devonian; and Irati - Permian) being responsible for generating hydrocarbons. In both systems, the maturation of organic matter, fluid migration and hydrocarbon trapping are related to the emplacement of intrusive igneous bodies (dykes and sills), which date back to the Jurassic-Cretaceous period (Arthur & Soares, 2002).

As previously mentioned, the Ponta Grossa geological formation dates back to the Devonian period and is lithologically made up of shales, silty shales and locally carbonaceous siltstones; fossiliferous and micaceous, with intercalations of light grey sandstones. Geochemical characterisation of samples from this formation revealed total organic carbon (TOC) levels ranging from 1 to 3 %, with an average of 0.5 %; with a predominance of organic matter of marine origin (type II kerogen) in the western part of the basin, and of continental origin in the eastern part (type III kerogen). The Ponta Grossa formation is classified as a proven generator of natural gas and oil condensate; reaching super-mature levels in a large part of the basin. The hydrocarbon generating potential (S2) reaches values in the order of 6.6 mg of HC/g of rock, and low hydrogen index (HI) values, based on vitrinite reflectance analysis, which result from the formation's high stage of thermal maturation (Arthur & Soares, 2002) (Milani, 2004) (Azevedo da Silva, 2007).

The Irati formation, dating from the Permian, is made up of shales enriched in amorphous organic matter (favourable for the generation of liquid hydrocarbons). Geochemical analyses reveal that the total organic carbon (TOC) content in the formation ranges from 0.1 to 23%, with an average of 2.0%.

Its potential for hydrocarbon generation (S2) is between 100 and 200 mgHC/g of rock; the predominant organic matter is of algae origin (type I kerogen, with types II and III also present). In general, the Irati formation shows a low degree of thermal evolution, with the exception of the basin's depocentre (central rift), where it can reach high maturity values due to the influence of intrusive magmatic bodies (Arthur & Soares, 2002) (Milani, 2004) (Azevedo da Silva, 2007).

Based on this information, the potential for generating unconventional hydrocarbons in the Irati and Ponta Grossa formations, in the Paraná Basin, can be assessed, as it coincides with the potential for generating conventional hydrocarbons but does not require the presence of traps to accumulate oil: the low permeability characteristic of the lithology - shales - acts as a reservoir and sealant for the petroleum system, characterising it as unconventional.

10.4. Geological characterisation of the Irati and Ponta Grossa Formations

According to studies carried out by Azevedo da Silva (2007), it can be concluded that the Irati and Ponta Grossa geological formations are the ones with the potential for hydrocarbons in the Paraná Basin. Based on geochemical analyses, the author summarised the data obtained from boreholes and samples from outcrops throughout the basin (Figure 42) (Table 9) and assessed the potential for hydrocarbon generation in these formations.

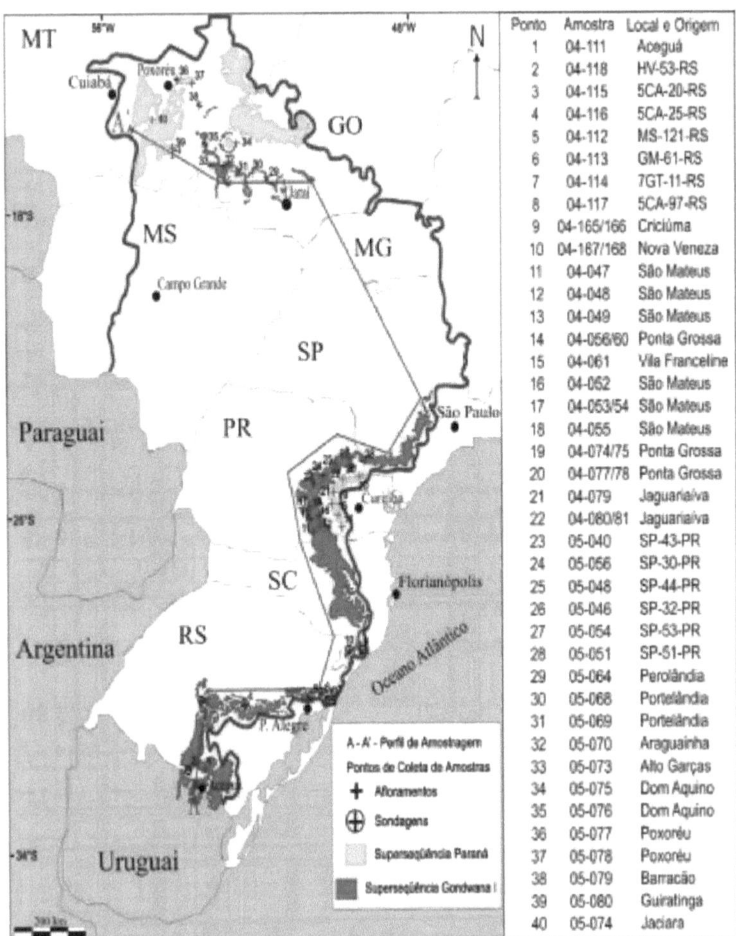

Figure 42 - Map showing the location of the exploratory wells in the Paraná Basin and the points sampled in the Irati and Ponta Grossa formations. Source: Azevedo da Silva (2007).

Table 9 - Results obtained from analyses of TOC, Rock-Eval pyrolysis in samples from the Irati and Ponta Grossa formations, Paraná Basin.

identification	Training	TOC (%)	S1	S2	S1+S2	S3	Tmax	HI	OI	IP	Ro (%)
04-056	Ponta Grossa	1.81	0.34	7,97	8.31	0,06	444	440.33	3.31	0.04	0,54
04-074	Ponta Grossa	1.19	0.12	1,61	1.73	0,60	437	135,29	50.42	0,07	0,55
04-077	Ponta Grossa	1.12	0.08	2,55	2.63	0,11	445	227.68	9.82	0,03	0.48

04-079	Ponta Grossa	1.00	0.13	0,01	0.14	0,23	437	1,00	23.00	0,93	0,53
05-077	Ponta Grossa	2,53	0.97	6,20	7.17	0,34	427	225.45	12.36	0,14	0,55
04-049	Irati	21,00	4.91	131.16	136,07	8,02	419	624.57	38,19	0,04	0,51
04-055	Irati	13,40	5.73	99,87	105.60	4,76	424	745,30	35,52	0,05	0,51
05-079b	Irati	3.31	1.58	10,11	11,69	0.65	388	368,98	23.72	0,14	0,56
05-064	Irati	2.41	8.41	63,13	71,54	0,04	337	756,45	0.98	0,12	0,45
04-165	Irati	5.49	1.39	66,27	67,66	0,42	432	726,64	4,60	0,02	0,52
04-166	Irati	1.28	1.66	51,55	53,21	0,27	429	991.34	5,19	0,03	0,47
04-117	Irati	1.34	0.14	1,24	1.38	0,43	431	92,54	32,08	0,10	0,58
04-118	Irati	1.68	0.24	2,85	3.09	0,62	417	169.64	36.98	0,08	0,56

Source: Azevedo da Silva (2007).

Based on this data analysis, it can be seen that for the samples analysed, the Irati formation has a high organic matter content with medium to high conditions for preserving this organic matter, while the Ponta Grossa formation has low to medium preservation conditions. The influence of basaltic intrusions on rock maturation can also be observed: samples collected in the vicinity of igneous intrusive rocks have low TOC levels, a condition resulting from the thermal cracking of the organic matter present (Azevedo da Silva, 2007). The author also identified the predominant type of kerogen in the samples: type I in the Irati formation and type II in the Ponta Grossa formation. The Tmax (temperature at which maximum hydrocarbon generation occurs) and IP (production index) values obtained through Rock-Eval pyrolysis showed a low degree of thermal evolution for samples from the Irati formation without the influence of intrusive igneous bodies, while samples from the Ponta Grossa formation show a high degree of thermal maturity (supermature rocks), as summarised in Table 10.

Table 10 - Application of the essential parameters for the occurrence of unconventional natural gas to the Irati and Ponta Grossa geological formations, Paraná Basin.

Comparative	Irati	Ponta Grossa

Parameters		
Geological Age	Permian	Devonian
Lithology	Black shales enriched in amorphous organic matter and with carbonate lenses; dolomitic limestones.	Laminated and carbonaceous black shales, silty shales, locally carbonaceous, fossiliferous and micaceous siltstones, with sandstone intercalations.
Mineralogical composition Predominant (% volume)	Silica (40-55), Argillominerals (4560)	**> Silica**
Mineralogy (% volume)	Mica and clay minerals (60-70), titanite and rutile (1-8), organic matter (20-30)	-
Area (km)²	-	
Thickness (m)	**<20**	80 (on average from 20 to 60)
Depth (m)	< 3.000	<3.500
Type of kerogen Predominate	I, II and III (types I and II predominate)	II and III (type II predominates in the western part of the basin)
Ro (%)	1,1-1,6	**> 1,35**
TOC (%)	**1-23 (average of <2)**	1-3 (average 0.5)
Tmax (°C)	**<440**	430 - 500
Thermal Maturity	Associated with the rate of burial: higher in the basin depocentre; and the influence of intrusive magmatic bodies. predominantly immature.	Associated with the rate of burial and the influence of intrusive magmatic bodies. predominantly supennatures (a domain more suited to the generation of unconventional natural gas).
Basin Classification Sedimentary		
Depositional palaeoenvironment		
Reservoir characteristics (%)	S1 = 0.02 - 0.8 mgHC/g rock; S2 = 100 - 200 mgHC/g rock; S1+S2 = 0.2 **-0.8; IP <0.1**	S2 = 6.6 mg; S1+S2 = potential for natural gas;

Based on the analysis of the data collected and the geological, geochemical and petrophysical parameters essential to the occurrence of unconventional natural gas (shale gas), based on fields proven to produce the resource and on studies by (Arthur & Soares, 2002) (Milani, 2004) (Azevedo da Silva, 2007); it can be concluded that:

- The Irati Formation has moderate potential for generating unconventional natural gas, and this potential is concentrated in deeper areas of the basin (depocentre). The Irati Formation is at a low stage of thermal maturation, and it is necessary to find areas where the formation is thicker and closer to sills (horizontally arranged intrusive igneous bodies whose geothermal gradient favours the maturation of organic matter), and away from dykes and faults (sudden increase in temperature and consequent cracking of organic matter and loss of the hydrocarbon generated);

- The Ponta Grossa Formation also has moderate potential for generating unconventional natural gas. However, it has comparatively greater potential than the Irati Formation due to its greater stage of thermal evolution, greater thickness and depth. However, it is necessary to find areas that are richer in organic matter (sweet spots) and with a higher degree of organic matter preservation (areas far from intrusive igneous bodies), conditions that are present in the westernmost part of the basin and extend into Uruguay (Arthur & Soares, 2002).

According to studies by Amaral (2016), the Irati formation in the state of Rio Grande do Sul, where the general direction of the gondwanic layers is E-W, dips to the north in line with the general configuration of the Paraná basin. However, analyses of boreholes in the São Gabriel region (extreme south of the state) indicate the beginning of the inflection of the layers to the south, i.e. the direction of the Irati becomes NE with a dip to the NW in the southernmost part of the basin. Although no data was found on the direction and inclination of the Irati in Uruguay, the formation is described in the literature as being lenticular in character and constantly present; although a thickness of 70 metres is described, which is almost double that of the Irati in Brazil (where it is between 20 and 40 metres thick), indicating that it is a continuous formation. Prospecting work carried out by American companies since 2009 in Uruguayan territory points to evidence of hydrocarbons, mainly natural gas in the La Paloma, Durazno region (Cordobés formation); at a depth of around 180 metres. Exploration of the resource began in 2011, as a result of a partnership between the state company ANCAP and Venezuela's PDVSA (Amaral, 2016).

In Uruguay, the Devonian sequence also shows a transgressive-regressive cycle, lithologically represented by the Durazno Group (Bossi, 1966 apud Oliveira, 1997), a name initially attributed to outcropping sediments in the central region of Uruguay, with an area of approximately 2,200 km^2 . This group includes the following geological formations: Cerrezuelo (sandstones), Cordobés (black shales) and La Paloma (sandstones), which form an oil system (Oliveira, 1997). The subsurface sedimentary package of the Durazno Group corresponds to thicknesses varying between 190 and 280 metres, delimited both at the base and at the top by regionally significant stratigraphic disagreements.

The Cordobés Formation is the formation belonging to the Durazno Group that is most relevant to this study due to its correlation with the Ponta Grossa Formation of the Paraná Basin. According to Oliveira (1997), the Cordobés formation is lithologically made up of dark grey to black, very fossiliferous, micaceous shales and silty shales, with significant occurrences of pyrite. There are also thin intercalations of grey, micaceous, fine-grained sandstones, the result of the gradual contact between the latter and the overlying La Paloma geological formation. Analyses of stratigraphic wells in the La Paloma region, in central Uruguay, reveal a maximum thickness of 117 metres for the Cordobés formation.

CHAPTER 12

HISTORY OF OIL EXPLORATION IN THE BASIN

Research into oil deposits in the Paraná Basin began in Brazil in 1892, in the region of Bofete, in the state of São Paulo, by Eugênio Ferreira de Camargo, who drilled wells at an average depth of 500 metres, which only produced sulphurous water. Later, at the beginning of the 20th century, in 1901 and 1902, new research was carried out by Companhia Cruzeiro do Sul, headed by Arthur B. Reardon. Reardon, also without economically viable results. In 1906, the Government of the State of São Paulo carried out the first official drilling of exploratory wells, near the town of Guarei; represented by Governor Jorge Tibiriçá and the Geography and Geology Commission of the State of São Paulo. Although no significant concentrations of hydrocarbons were found, the sampling resulted in knowledge of the lithological profile present in the region (Galhano, 2006).

The first federal oil prospecting initiatives in Brazil began in 1919 through the Geological and Mineralogical Service of Brazil, initially drilling in the states of Paraná and São Paulo under the command of President Delfim Moreira. In 1921, commercial production of natural gas began in the basin through wells around 310 metres deep (Corumbataí Formation). In 1932, the Companhia de Petróleo do Brasil, headed by Monteiro Lobato, drilled into the basalt layer in the sedimentary basin (around 1,076 metres deep), but without much economic return. In 1946, the Sociedade Brasileira de Sondagens was organised, under the technical direction of Silvio Fróes de Abreu, which carried out one of the most extensive hydrocarbon exploration programmes in the Paraná Basin, but which also ended its activities in 1950 due to lack of economic return. From 1953, with Law 2004, hydrocarbon exploration in Brazil became a federal monopoly, and was exclusively carried out by Petrobras (Yoshida, 1978 apud Galhano, 2006).

From 1953 to 1974, Petrobras made a major systematic effort to explore for hydrocarbons in the Paraná Basin: 71 exploratory wells were drilled and different prospecting methods were applied, generally of an experimental nature, such as aeromagnetometry, gravimetry and reflection seismic; activities which contributed to the collection of information about the basin. However, these studies were geographically concentrated in a strip of land equivalent to just 30% of the total area of the sedimentary basin, and obtained unpromising results in reflection seismic due to the presence of basaltic layers in the geological formation, which cover approximately 70% of the basin, making it difficult to interpret the seismic data. In addition to the difficulties associated with the application of geophysical techniques, the basaltic cover would have contributed to thermal cracking in potentially hydrocarbon-generating rocks, promoting their oversaturation and the consequent degradation of hydrocarbons; as well as the weak structuring of the basin, which implies a reduction in accumulating structures, factors evidenced in the absence of significant oil accumulations in the region (Galhano, 2006). Subsequently, and based on the development of further studies into the geology of the Paraná Basin, it was concluded that, despite the factors limiting hydrocarbon accumulation described above, the basin had lithology favourable to the generation of oil reservoirs, which could be exploited using unconventional exploration and production mechanisms, the focus of this study.

CHAPTER 13

BRAZILIAN ENERGY SCENARIO

13.1. Uruguaiana Thermoelectric Plant

AES Uruguaiana began operations in 2000 in the city of Uruguaiana, in the state of Rio Grande do Sul, under an authorisation to operate as an independent electricity producer for a period of 30 years (from 26 June 1997). The plant has an installed capacity of 639.9 megawatts (MW) and has two combustion turbines with a capacity of 187.5 MW each and a steam turbine with a capacity of 265 MW. The plant is the first thermoelectric plant to run on natural gas in Brazil, a comparatively less polluting fuel compared to the other fossil fuels used for thermoelectricity: coal and diesel (AES Uruguaiana S.A., 2014).

The thermoelectric plant's activities have been paralysed since April 2009 due to the total interruption in the supply of natural gas by its sole supplier, the Argentine company Yacimientos Petrolíferos Y Fiscales S.A. (YPF), since May 2008; an interruption which has led to the plant's operational activities being paralysed since that date. Also in 2009, AES Uruguaiana filed a request for arbitration against YPF over the breach of the supply contract, seeking payment of compensation. A favourable decision was handed down in May 2013, but the dispute is still under discussion in the arbitration court and is expected to be concluded in 2017. At the end of 2012, the Ministry of Mines and Energy (MME), together with AES Uruguaiana, Petróleo Brasileiro S.A. (Petrobras), Companhia de Gás do Estado do Rio Grande do Sul (Sulgás), Transportadora Sulbrasileira de Gás (TSB) and the National System Operator (ONS), began negotiations aimed at restarting the plant for a period of 60 days. Operations were partially resumed on 6 February 2013 (initially with a capacity of 164 MW and later with a combined cycle of 244 MW). In 2014, the plant was again authorised by the MME to operate on an exceptional and temporary basis, generating around 240 MW during the period from 9 March to 16 May 2014 (AES Uruguaiana S.A., 2014).

Since the interruption in the plant's fuel supply (natural gas) in April 2009, the plant has stopped its activities and started a maintenance and conservation programme (hibernation); a decision taken with the aim of reducing costs and preparing the plant for the possible resumption of activities in the future. Since the MME recognised the plant's need for exceptional and temporary generation between 2013 and 2014, it recorded net generation of 322.1 GW in 2014, 35% more than in the previous year (239.0 GW). The plant's resumption of operation during this period required investments from AES Uruguaiana to bring it out of hibernation, and this amount was reimbursed through System Service Charges (ESS) totalling R$22.5 million, recorded by the Electricity Trading Chamber (CCEE) in September 2013. In 2014, R$259 million was reimbursed in the CCEE settlement process, and R$3 million was also requested from the National Electricity Agency (ANEEL) to cover additional costs associated with the process. Overall, the plant's operational return during 2013 and 2014 resulted in gross revenue of R$265.5 million in 2014 and R$188.9 million in 2013. This rise in costs was related to factors such as the increase in power generation (from 188.9 GW in 2013 to 322.1 GW in 2014) and the increase in the price of the required fuel, natural gas, also recorded in the period: the lack of input of the resource triggered the price increase, in addition to the consequent need to import liquefied natural gas (LNG), also associated with higher costs. The import of LNG is brokered by Petrobras (fuel imported from Nigeria and Trinidad Tobago), after which the LNG is delivered to Sulgás, Rio Grande do Sul's natural gas distributor, which contracts the regasification and transport of the energy source in Argentina and then in Brazil to supply the thermal power station. In this context of higher gas consumption and higher prices, the company's operating costs and expenses, totalling R$171.2 million, correspond to an increase of 44% between 2013 and 2014 (AES Uruguaiana S.A., 2014).

CHAPTER 14

COMPARATIVE ANALYSIS

Table 11 - Comparative analysis of the parameters essential to the occurrence of unconventional natural gas between the Marcellus, Barnett, Eagle Ford, Irati and Ponta Grossa geological formations.

Parameters Comparisons	Marcellus	Barnett	Eagle Ford	Irati	Ponta Grossa
Age Geological	Middle Devonian	Upper Mississippian	Upper Cretaceous	Lower Permian	Middle Devonian
Lithology	Black leaflets carbonaceous, highly radioactive carbonate concretions ; shales enriched in silica and organic matter.	Dense, siliceous , oil-bearing black shales , fossiliferous and rich in organic matter carbonates crystalline, oil-bearing and fossiliferous; black shales highly enriched in silica (more rutile).	Carbon leaflets rich in material organic; carbonates oil and fossil fuels also enriched in organic matter.	Black leaflets enriched in amorphous organic matter and with lentescarbonates ; dolomitic limestone.	Black leaflets laminated and carbonaceous, silty shales, locally carbonaceous, fossiliferous and micaceous siltstones, with interbeds sandstone.
Predominant Mineralogical Composition (% by volume)	Silica (10-20%); clay minerals (35%); carbonates (50%)	Silica (35-50%); clay minerals (<35%); carbonates (20%)	Silica (15-20%); clay minerals (15-30%); carbonates (40-90%)	Silica (40-55%), Argillominerals (45-60%)	>Silica (compared to Irati)
Mineralogy (% volume)	27-31% quartz, 9-34% illite, 1-7% clay mineral complex, 0-4%	30-50% quartz, 27% illite and traces of smectite, 8-19% calcite and dolomite,	15-20% quartz, 15-30% clay minerals, 40-90% calcite and dolomite +	Mica and clay minerals (60-70%), titanite and rutile (1-8%),	-

	chlorite, 3-48% calcite, 0-10% dolomite, 0-4% sodium feldspar, 5-13% pyrite, and 0-6% gypsum.	7% feldspar, 5% organic matter, 5% pyrite, 3% siderite, and traces of phosphate and copper.	organic matter + pyrite.	organic matter (20-30%)	
Area (km)2	60.000	13.000	12.140	770,000 (around 70 per cent of the basin's total area)	1,100,000 (the entire length of the basin in Brazil) + continuity in Uruguay (Durazno Group)
Thickness (m)	15-200 (typically 1580)	15-300 (typically 45-180).	15-150 (typically 70)	20 -40 (typically < 20)	80 (typically from 20 to 60)
Depth (m)	600-3.000	1.200-2.500	1.200-4.000	<3.000	<3.500
Type Predominate kerogen	II and III (predominant type II)	II (only type III traits)	II and III	I, II and III (types I and II predominate)	II and III (type II predominates in the western part of the basin)
Ro (%)	1.60 (maximum 3.5)	1.2 (maximum 1.9)	1,4	1,1 - 1,6	> 1,35
TOC (%)	1-10	2-6	2-12	1-23 (average < 2)	1-3 (average < 2)
Tmax (°C)	470	465	335	<440	430 - 500
Maturity Thermal	Predominantly supermature rocks (gas thermogenic). Associated with the depth of burial and	Rocks that vary from immature to supermature. Associated with the depth of burial and hydrothermal activity along fault and fracture	Rocks that vary from immature to supermature. Associated with the depth of burial and hydrothermal activity	Associated with the rate of burial: higher in the basin's depocentre; and the influence of intrusive magmatic	Associated with the burial rate and the influence of intrusive magmatic bodies. Predominantly supermatur

	thermal flow in fault zones.	zones.	along fault and fracture zones.	bodies. Predominantly immature rocks.	e rocks (the most suitable area for generating unconventional natural gas).
Basin Classification Sedimentary	Retro-arc, asymmetrical, delimited by orogenic belts , arcs structure pre-construction Cambrian.	Retro-arch, asymmetrical, bounded by prominent orogenic belts, structural arcs and Precambrian uplift.	Retro-arc, asymmetrical, delimited by orogenic belts.	Intracratonic. Structural boundaries. Low complexity structural.	Intracratonic. Structural boundaries. Low complexity structural.
Palaeoenvironment Depositional	Deep-water sedimentary basin, low circulation, stratified water column, reducing environment at the base, presence of algae-planktonic organic matter, water depth around 100 to 300 metres.	Deep-water sedimentary basin, low circulation, stratified water column, reducing environment at the base, presence of algae-planktonic organic matter, water depth of around 150 to 200 metres.	Sedimentary basin on the offshore shelf-continental circulation, stratified water column, reducing environment at the base, presence of algae-plankton organic matter, water depth of around 100 metres.	Deep-water sedimentary basin, low circulation, stratified water column, reducing environment at the base, presence of algae-planktonic organic matter,	Deep-water sedimentary basin, low circulation, stratified water column, reducing environment at the base, presence of organic matter
Characteristics Reservoir (%)	φ= 6, k = micro-nanodarcies, Sw = 30 (no free water), Sg = 70 (gas allocated to	φ = 3-6, k = micro-nanodarcies, Sw = 25 (no free water), Sg = 75 (gas allocated to interstitial	φ = 8, k = micro-nanodarcies, Sw = 20 (no free water), Sg = 75 (gas allocated to	S1 = 0.02 - 0.8 mgHC/g rock; S2 = 100 - 200 mgHC/g rock; S1+S2 = 0.2 -	S2 = 6.6 mg; S1+S2 = potential for natural gas;

	interstitial pores) and Sg = 70 (gas allocated to inters titial pores). microfractur es, and adsorbed to organic matter and kerogen: 20-80%); internal reservoir pressure: from 0.51 psi/ft (overpressu re) to 0.10-0.35 psi/ft (low pressure)	pores and microfractures , and adsorbed to organic matter and kerogen: 20-60%); overpressure stage: 0.52psi/ft	interst itial pores), Sg = 10 (no free water). microfractur es, and adsorbed to organic matter and kerogen: 20-80%)internal reservoir pressure: 0.50.8 psi/ft, average 0.65 psi/ft (overpressu re)	0.8;IP<0.l	

CHAPTER 15

FINAL CONSIDERATIONS AND CONCLUSIONS

With a view to establishing and developing a national natural gas industry, proving the potential for large continental gas accumulations in oil systems with an unconventional geological configuration (unconventional natural gas) is of fundamental importance for the Brazilian energy sector, as well as highlighting the contribution of geological knowledge to energy planning. From an economic point of view, current natural gas production in Brazil essentially comes from conventional reserves, which are concentrated in remote areas (offshore sedimentary basins, or in the Amazon region) and the consequent distance from the main consumer centres (south and southeast regions of Brazil) requires high investment in distribution infrastructure and higher associated costs. In addition to its economic importance and the potential reduction in dependence on natural gas imports and consequent reduction in costs, natural gas has advantages over other energy sources in terms of environmental aspects: comparatively low CO_2 emissions compared to other fossil fuels (coal and fuel oil), as well as a lower degree of uncertainty and risk compared to nuclear energy.

Given this scenario, the possibility of producing natural gas from unconventional reservoirs, which are widely distributed in Brazil's sedimentary basins, through the implementation of specific exploratory techniques (horizontal drilling and hydraulic fracturing), could encourage Brazilian production of unconventional natural gas, making this resource more competitive in the oil market and potentially changing the national energy landscape (Tassinari, 2015).

In the context of Brazilian sedimentary basins with the potential to generate unconventional natural gas (shale gas), the Paraná sedimentary basin stands out due to its high potential for this resource, as presented in this study (favourable geological configuration, potential for generation and storage in the Irati and Ponta Grossa formations, geochemical parameters corresponding to those found in the main shale gas producing fields today: Barnett, Marcellus and Eagle Ford in the

United States); and its proximity to the main energy consuming centres in Brazil, covering the states of SP, PR, SC, RS, GO,MT, MS and MG. In addition, the concentration of electricity production in Brazil on its hydroelectric potential makes the Brazilian energy matrix relatively vulnerable to climatic-seasonal variations, such as the 2014 water crisis, which significantly impacted hydroelectric projects and triggered an increase in the cost of energy generation, passed on to the population.

Based on the data presented in the comparative table above, the potential for the occurrence of unconventional natural gas in the Irati and Ponta Grossa formations in the Paraná Basin in Brazil can be inferred. This potential is comparatively established according to data from formations that have been proven to produce this resource (Barnett, Marcellus and Eagle Ford in the United States).

Considering the geological age of the formations analysed, the Brazilian formations date from periods favourable to the generation of hydrocarbons and consistent with the time intervals in which the North American formations fall (predominantly between the Devonian and Permian, average 391.8 to 275.6 Ma).

Based on the lithology of the formations, they all show compositional similarities: made up of black shales rich in organic matter and carbonaceous fractions, with variations in silica and clay concentrations. Considering this variation in the predominant mineralogical composition, both Brazilian formations (Irati and Ponta Grossa) are favourable for the implementation of the exploratory technique of hydraulic billing due to the high silica content in their compositions, which must be greater than 30% of the total volume with carbonate fractions present, according to Andrews (2013), a characteristic that gives the shale a rutile aspect and the consequent greater effectiveness in its induced billing. Compared to each other, the Brazilian formations differ mineralogically in their carbonaceous and siliceous fractions, with the Ponta Grossa Formation being more favourable for implementing the aforementioned technique due to its higher silica and carbonate content.

In terms of area of occurrence, both Brazilian formations are large enough for economically viable hydrocarbon exploration, with the Ponta Grossa formation being the largest in km^2 , extending throughout the Paraná Basin as far as Uruguay, where it is thicker and deeper. Analysing these parameters, the Brazilian formations are at a relative disadvantage compared to the American formations, which have greater thicknesses and depths and, consequently, greater potential for generating natural gas. However, parameters such as high organic matter content and thermal maturity may compensate for this disadvantage: the presence of intrusive igneous bodies in the Paraná Basin possibly contributes to the thermal maturation process of these formations. In addition, the possibility of generating natural gas of biogenic origin could change this scenario, which is independent of the high rate of maturation, thickness and depth, but is related to the bacteriogenic degradation of organic matter (Tassinari, 2015). Still according to these aspects, the Brazilian formations still fit the parameters considered favourable for the generation of unconventional natural gas, mentioned previously: thickness necessarily greater than 15 metres and depth greater than 1,500 metres. In comparison, the Ponta Grossa Formation is thicker and deeper than the Irati Formation, with progressively higher values towards the south of the Paraná basin, in the state of Rio Grande do Sul, and in Uruguai, the areas closest to the study site (Uruguaiana).

Considering the tectonic and diagenetic aspects of the sedimentary basins analysed, they are similar: generally intracratonic, of low structural complexity and formed in a deep-water palaeoenvironment with low water circulation and a reducing environment rich in organic matter.

According to Andrews (2013), the most suitable total organic carbon (TOC) values for unconventional natural gas in shales are higher than 2%. The most suitable type of organic matter is type II kerogen, with type I also being favourable (properties that favour the Irati Formation). Type I kerogen is more favourable for generating oil (oil prone); although it can generate gas, but in smaller quantities than type II (gas prone). Ideal vitrinite reflectance values vary between 1.1 and 3.5 % (preferably between 1.75 and 3.0 %), the margin where both the Irati and Ponta Grossa formations are found.

Based on the comparative analysis carried out, it can be concluded that both formations belonging to the Paraná Basin (Irati and Ponta Grossa Formations) have moderate to high potential for the generation and storage of unconventional natural gas, with the Ponta Grossa Formation having the greatest potential for generation and storage and, consequently, being the most recommended for production of this resource in southern Brazil.

CHAPTER 16

BIBLIOGRAPHICAL REFERENCES

AES BRASIL. AES Uruguaiana. Available at <http://www.aesbrasil.com.br/Paginas/aes-brasil.aspx> Accessed 08/2015.

OIL, NATURAL GAS AND BIOFUELS (ANP). Available at < http://www.anp.gov.br/?id=2803>. Accessed 08/2015.

ANDERSON, J., SIMPSON, M., BASINSKI, P., BEATON, A., BOYER, C., Producing Natural Gas from Coal. Oil Field Review. 2003.

ANDREWS, I.J. 2013. The Carboniferous Bowland Shale gas study: geology and resource estimation. British Geological Survey for Department of Energy and Climate Change, London, UK.

AMARAL, Sérgio Estanislau do. Geology and petrology of the Irati Formation (Permian) in the state of São Paulo. Bol. IGA, São Paulo, 2016. Available at <http://ppegeo.igc.usp.br/scielo.php?script=sci_arttext&pid=S01003925197100 0100001&lng=en&nrm=iso>. Accessed on 29 January 2016.

ARTHUR, P.C., SOARES, P.C. 2002. Brazilian Journal of Geosciences. 32 (4), 2002, pp.433-448.

AZEVEDO DA SILVA, C.G. 2007. Organic geochemical characterisation of oil-bearing rocks from the Irati and Ponta Grossa formations of the Paraná Basin. Master's dissertation. Chemistry Institute, Federal University of Rio Grande do Sul. Porto Alegre, December 2007.

BARBOSA, H. C. R., CASTRO, A. B. de, CUPERTINO, D. F., DIAS, K. D. N., PECCINI, T. L. Structural geology, tectonics and sedimentation of the intracratonic basins: Solimões, Amazonas, Parnaíba and Paraná. 142 f. Final Course Work (Specialisation Course in Basin Analysis Project: Petroleum Geology Module) Faculty of Geology, Rio de Janeiro State University, Rio de Janeiro, 2007.

BOYCE, M. Subsurface stratigraphy and petrophysical analysis of the Middle Devonian interval of central Appalachian basin: West Virginia and southwest Pennsylvania: PhD thesis proposal. 2009. West Virginia University, 23 p.

BOWKER, K.A., Recent developments of the Barnett shale play, Fort Worth basin, inLaw, B.E., and Wilson, M., eds., Innovative gas exploration concepts symposium: Rocky Mountain Association Geologists and Petroleum Technology Transfer Council, 2002. Oct 2002, 16 p.

BRUNER, K.R., SMOSNA, R., A Comparative Study of the Mississippian Barnett

Shale, Fort Worth Basin, and Devonian Marcellus Shale, Appalachian Basin. U.S. Department of Energy. 2011. DOE/NETL-2011/1478.

CHAPMAN, R.E., Petroleum Geology: A Concise Study. Elsevier Scientific Publishing Company. Amsterdam, Oxford, New York. 1976.

CHARPENTIER, R.R. & COOK, T.A., 2011. USGS methodology for assessing continuous petroleum resources. USGS Open-File Report. 2011-1167.

COMPANHIA DE GÁS DE SÃO PAULO. Natural gas in industry.

Available in <http://www.comgas.com.br/quero_industria/gasnatural/segmentos.asp> Accessed 08/2015.

GAS COMPANY OF THE STATE OF RIO GRANE DO SUL. Available at <http://www.sulgas.rs.gov.br/sulgas> Accessed 08/2015.

DAS VIRGENS, G.B., Bibliographic Review of the Barnett Formation Gas Sheets, Texas, USA: an example of an unconventional reservoir. Final Undergraduate Paper. Federal University of Bahia. Institute of Geosciences. p.85. 2011.

DEUTSCHE WELLE. Environment. Available at < http://www.dw.de/brasil-quer-usar-fraturamento-hidr%C3%A1ulico-para- explorar-g%C3%A1s-de-xisto/a-16794512> Accessed 03/2014.

DURHAM, L.S. Appalachian basiifs Marcellus-the new target: AAPG Explorer, March2008 : http://www.aapg.org/explorer/2008/03mar/marcellus.cfm, accessed July 2009.

DURHAM, L.S., Barnett Shale play still going strong: American Association Petroleum Geologists Bulletin, 2005, v. 86, p. 1,921-1,938.

ENERGY INFORMATION ADMINISTRATION, EIA. Available at <http://www.eia.gov/energy_in_brief/article/about_shale_gas.cfm> Accessed 08/2015.

ENGELDER, T., Structural geology of the Marcellus and other Devonian gas shales: geological conundrums involving joints, layer-parallel shortening strain, and the contemporary tectonic stress field: Pittsburgh Association of Petroleum Geologists Field Trip. 2008a. (Sept. 12-13, 2008), 96 p.

ENGELDER, T., and Lash, G.G., Marcellus Shale play's vast resource potential creating stir in Appalachia: The American Oil and Gas Reporter, May 2008, 7 p.

EOFF, J.D., Global Predition of Continuous Hydrocarbon Accumulations in Self-Sourced Reservoirs. U.S. Geological Survey Open-File Report 20121091. 2012.

EXXONMOBIL. About Natural Gas. 2012. Available at <http://aboutnaturalgas.com/content/natural-gas/tight-and-shale-gas/> Accessed 03/2014.

Available at <http://www.forbes.com/sites/energysource/2012/12/07/surprise-side-effect-of-shale-gas-boom-a-plunge-in-u-s-greenhouse-gas-emissions/> Accessed 03/2014.

FRACFOCUS. Chemical Disclosure Registry. Hydraulic Fracturing: The Process. 2010. Available at < http://fracfocus.org/hydraulic-fracturing-how-it-works/hydraulic-fracturing-process> Accessed 03/2014.

FUGITA, A.M. Fourth Bid Round. Parnaíba Basin. Block Definition Superintendence. ANP - National Agency for Petroleum, Natural Gas and Biofuels. Rio de Janeiro, 2004.

HALLIBURTON. Coalbed Methane: Principles and Practices. 2007. Available at http://www.halliburton.com/public/pe/contents/Books_and_Catalogs/web/CBM/H06263_Chap_01.pdf> Accessed 11/2014.

HARPER, J.A. The Marcellus Shale-an old "new" gas reservoir in Pennsylvania: Pennsylvania Geology. 2008. v. 38, p. 2-13: http://www.dcnr.state.pa.us/topogeo/pub/pageolmag/pdfs/v38n1.pdf, accessed July 2009.

HENTZ, T.F., and RUPPEL, Regional Stratigraphic and Rock Characteristics of Eagle Ford Shale in Its Play Area: Maverick Basin to East Texas Basin, Search andD iscovery Article#10325,(http://www.searchanddiscovery.com/pdfz/documents/2011/10325hentz/ndx_hentz.pdf.html) based on oral presentation at AAPG Annual Convention and Exhibition, Houston, Texas, USA, April 10-13, 2011.

HUBBERT, M.K., Energy Resources. National Academy of Sciences, Publication 1000-D, p.81-83. 1962.

KHAMSIN, S.A., Basic Properties of Reservoir Rocks. Department of Petroleum Engineering. King Fahd University of Petroleum and Minerals. Dhahran, Saudi Arabia, 2004.

LANCASTER, D.E., McKetta, S., and Lowry, P.H., Research findings help characterise Fort Worth Basin's Bamett Shale: Oil & Gas Journal, 1993, v. 91, issue 10, p. 59-64.

LAW, B.E., Basin Centred Gas Systems. AAPG Bulletin, v.86, no.11. P.1891-1919. 2002.

LEE, D.S., Herman J.D., Elsworth, D., Kim, H., Lee H. A critical evaluation of unconventional gas recovery from the Marcellus shale, northeastern United States. KSCE. Journal of Civil Engineering. 2011. 15 (4):679-687.

LINK, P.K. Basic Petroleum Geology. Oil and Gas Consultants International (OGCI). Tulsa, OK. 3rd edition. p.232. 2007.

LIMA GOMES, A.J. 2009. Evaluation of geothermal resources in the Paraná Basin. Doctoral thesis. MCT National Observatory. Coordination of Postgraduate Studies in Geophysics. Rio de Janeiro. ON, 2010.

LOUCKS, R.G., and RUPPEL, S.C., Mississippian Barnett Shale: lithofacies and depositional setting of a deep-water shale-gas succession in the Fort Worth Basin, Texas: American Association Petroleum Geologists Bulletin, 2007, v. 91, p. 579-601.

MAACK, R. 1947. Brief news on the geology of the states of Paraná and Santa Catarina. Archives of Biology and Technology. Curitiba, v.2, p. 63-154, 1947.

MAGOON, L.B. ; DOW, W.G. The Petroleum Systems: from source to trap. AAPG Memoir 60. Chapter 1. 1994.

MALAGUETA, D.C., The Environmental Dimension in the Delimitation of Unconventional Oil Reserves: The Cases of Canada and the Arctic. UFRJ/COPPE, 2009. XV, p. 110. Rio de Janeiro, 2009.

MATSUTSUYU, B. South Texas Eagle Ford Shale Geology- Regional Trends, Recent Learnings, Future Challenges. Momentum Oil & Gas. Developing Unconventional Gas Conference. Huston, 2011. Texas. October 10-12, 2011.

McCULLAGH, T., HART, B., Stratigraphic controls on production from a basin-centred gas system: Lower Cretaceous Cadotte Member, Deep Basin, Alberta, Canada. AAPG Bulletin, v.94, no.3. p.293-315. 2010.

MENEZES, T.R., FILHO, J.G.M., ARAUJO, C.V., SOUZA, I.V., MENDONÇA, J.O., Organic facies: concepts, methods and case studies in the petroleum industry. Brazilian Journal of Geosciences. 80-96 June 2008.

MILANI, E.J., BRANDÃO, J.A.S.L., GAMBOA, L.A.P. Petróleo na Margem Continetal Brasileira: Geologia, Exploração, Resultados e Perspectivas. Brazilian Journal of Geophysics, V. 18(3), 2001.

MILANI, E.J. et al. 2007. Paraná Basin. Bol. Geoc. Petrobras 15(2):265- 287.

MILANI, E. J. Tectonostratigraphic evolution of the Paraná Basin and its relationship with the Phanerozoic geodynamics of south-western Gondwana.

1997. 2 v. Thesis (Doctorate) - Federal University of Rio Grande do Sul, Porto Alegre, 1997.

MONTGOMERY, S.L., Jarvie, D.M., Bowker, K.A., and Pollastro, R.M., Mississippian Barnett Shale, Fort Worth basin, north-central Texas: Gas-shale play with multi-million cubic foot potential: American Association Petroleum Geologists Bulletin, 2005, v. 89, p. 155-175.

NORTON ROSE FULBRIGHT, 2013. Shale Gas Handbook: a quick-reference guide for companies involved in the exploitation of unconventional gas resources. Norton Rose Fulbright (NRF) 11/13 UK.

OLIVERIA S.F. Palynology of the Devonian sequence of the Paraná Basin in Brazil, Paraguay and Uruguay: biochronostratigraphic, palaeoenvironmental and palaeogeographic implications. PhD Thesis. Institute of Geosciences. University of São Paulo. São Paulo, 1997.

PARRA, P.A., Rubio, N., Ramirez, C., Guerra, V.A., Campos, I.R., Trejo, M.D., Olguin, J. Vargas, C.H., Valbuena, R., Soler, D.F., Weimann, M.I., Lujan, V., Bonningue, P., Reyes, P.G., Martinez, R., Munoz, R., Rodriquez, E., and Garcia, M., 2013, Unconventional Reservoir Development in Mexico: Lessons Learned From the Frist Exploratory Wells, Society of Petroleum Engineers, 2013. Unconventional Resources Conference- USA held in The Woodlands, Texas, USA, 10-12 April 2013, SPE 164545.

PEREIRA, E., CARNEIRO, C.D., BERGAMASCHI, S., ALMEIDA, F.F.M., Evolution of the Palaeozoic Synclines: Solimóes, Amazonas, Parnaíba and Paraná Provinces. Geology of Brazil. Chap.V.p.374-418. 2008.

Available at < http://www.petronoticias.com.br/archives/19096> Accessed 03/2014.

RADOVIC, L.R., Energy and Fuels in Society: Analysis of Bills and Media Report. McGraw-Hill Custom Publishing. 1997.

ROEN, J.B., Geology of the Devonian black shales of the Appalachian basin: Organic Geochemistry. 1984, v. 5, p. 241-254:http://www.eesi.psu.edu/news_events/EarthTalks/2009Spring/materials2009spr/Roen84BlkShDevorggeoch.pdf, accessed July 2009.

ROSTIROLLA, S.P., Uncertainty Analysis of Petroleum Systems. Brazilian Journal of Geosciences. 29(2):261-270. 1999.

SCHLUMBERGER. Tight Gas and Oil. 2013. Available at < http://www.slb.com/services/technical_challenges/unconventional_resources/tight_gas.aspx> Accessed 03/2014.

SEKITAN NO HON, H. Mengenal CBM (Coal Bed Methane). 2010. Available at < http://imambudiraharjo.wordpress.com/2010/01/19/mengenal-cbm-coal- bed-methane> Accessed 11/2014.

GEOLOGICAL SERVICE OF BRAZIL. CPRM. 2013. Available at < http://geobank.sa.cprm.gov.br/> Accessed on 03/2014.

SILVA, C.G.A., Organic Geochemical Characterisation of Oil-bearing Rocks from the Irati and Ponta Grossa Formations of the Paraná Basin. Master's dissertation. Chemistry Institute. Federal University of Rio Grande do Sul. Porto Alegre, 2007.

SOEDER, D.J., and Kappel, W.M., Water resources and natural gas production from the Marcellus Shale: U.S. Geological Survey. 2009. Fact Sheet 20093032, 6 p.

TASSINARI, C.C.G. 2015. Class Notes: Introduction to Unconventional Oil and Gas Exploration. Institute of Energy and Environment. University of São Paulo.

TASSINARI, C. Class Notes. Unconventional Oil and Natural Gas. Institute of Energy and Environment. University of São Paulo. 2015

TISSOT, B.P. WELTE, D.H. Petroleum Formation and Occurrence: A New Approach to Oil and Gas Exploration, xviii. p.538. Berlin, Heidelberg, New York: Springer-Verlag. ISBN 3 540 08698 6. 1978.

U.S. GEOLOGICAL SURVEY. Map of Assessed Shale Gas in the United States, 2012.Available at < http://pubs.usgs.gov/dds/dds-069/dds-069-z/DDS- 69-Z_pamphlet.pdf> Accessed 08/2015.

US ENERGY INFORMATION ADMINISTRATION (EIA), 2013, Technically Recoverable Shale Oil and Shale Gas Resources: An Assessment of 137 Shale Formations in 41 Countries Outside the United States. Available at: http://www.eia.gov/analysis/studies/worldshalegas/

U.S.ENERGY INFORMATION ADMINISTRATION (EIA). Updates to the EIA Eagle Ford play maps. U.S.Department of Energy. 2015 Washington DC. https://www.eia.gov/maps/pdf/eagleford122914.pdf

WERNER, A.K., VINK, S., WATT, K., Jagals, P. 2014. Environmental health impacts of unconventional natural gas development: A review of the current strength of evidence. Sciences of the Total Environment. 505 (2015) 11271141.

WHITE, I.C. 1908. Report on coal measures and associated rocks in southern

Brazil. Rio de Janeiro: Comissão das Minas de Carvão de Pedra do Brasil, 1908. 300 p.

WILLIAMS, J.H., The Marcellus Shale Play: geology, development and water resource impact mitigation. New York Water Science Centre. U.S. Geological Survey (USGS). 2012.

YERGIN, D. The Quest: Energy, Security, and the Remaking of the Modern World. Published by the Penguim Group. 2011, 2012.

YOSHIDA, R. 1978. History of oil research in Brazil and the state of São Paulo. São Paulo, 1978.

ZALÁN, P.V. et al. 1990. The Paraná Basin. In: Gabaglia G.P.R. & Milani E.J. (coords.) Origin and evolution of sedimentary basins. Petrobras Technical Bulletin. Técn. Petrobras, p. 135-168.

ZALÁN, P. V.; WOLFF, S.; ASTOLFI, M. A. M.; VIEIRA, I. S.; CONCEIÇÃO, J. C. J.; APPI, V. T.; SANTOS NETO, E. V.; CERQUEIRA, J. R.; MARQUES, A. The Paraná Basin, Brazil. In: LEIGHTON, M. W.; KOLATA, D. R.; OLTZ, D. F.; EIDEL, J. J. (Ed.). Interior cratonic basins. Tulsa: American Association of Petroleum Geologists, 1990. p. 681-708. (AAPG.Memoir, 51).

ZA ZA ENERGY Company Corporation, Presentation, 2013, Investor Presentation December 2014, p. 14. http://www.zazaenergy.com/

ZOU, C., ZHANG, G., YANG, Z., TAO, S., HOU, L., ZHU, R., YUAN, X., RAN, Q., LI, D., WANG, Z., Concepts, Characteristics, Potential and Technology of Unconventional Hydrocarbons: On Unconventional Petroleum Geology. Petroleum Exploration and Development. Research Paper. V.4 (4). P. 413-428. 2013.

Printed by Books on Demand GmbH, Norderstedt / Germany